Practical Handbook on MEAT SCIENCE AND TECHNOLOGY

Practical Handbook on MEAT SCIENCE AND TECHNOLOGY

Jhari Sahoo
Professor-cum-Head, M.V.Sc. (APT), Ph.D. (LPT), FNAVS

Davinder Kumar Sharma
Associate Professor, M.V.Sc. (VPH), Ph.D.

Manish Kumar Chatli
Associate Professor, M.V.Sc. (APT), Ph.D. (LPT)

Department of Livestock Products Technology
College of Veterinary Science
Guru Angad Dev Veterinary and Animal Sciences University
Ludhiana – 141 004 (Punjab)

2022

Daya Publishing House®
A Division of
Astral International Pvt. Ltd.
New Delhi - 110 002

First Published, 2011
Reprinted, 2022

ISBN: 978-81-7035-676-9 (HB)

Published by : **Daya Publishing House®**
A Division of
Astral International Pvt. Ltd.
– ISO 9001:2015 Certified Company –
4736/23, Ansari Road, Darya Ganj
New Delhi-110 002
Ph. 011-43549197, 23278134
E-mail: info@astralint.com
Website: www.astralint.com

Digitally Printed at : Replika Press Pvt. Ltd.

Tel. : 0161-2553360, 2553320
Fax : 0161-2553340
E-mail : vijay_taneja@hotmail.com
: vcgadvasu@hotmail.com

ਗੁਰੂ ਅੰਗਦ ਦੇਵ ਵੈਟਰਨਰੀ ਅਤੇ ਐਨੀਮਲ ਸਾਇੰਸਿਜ਼ ਯੂਨੀਵਰਸਿਟੀ
GURU ANGAD DEV
VETERINARY & ANIMAL SCIENCES UNIVERSITY
FEROZEPUR ROAD, LUDHIANA-141004 (INDIA)

ਡਾ. ਵਿਜੇ ਕੁਮਾਰ ਤਨੇਜਾ
ਉਪ-ਕੁਲਪਤੀ

Dr. Vijay Kumar Taneja
Vice-Chancellor

D. O. No.

Dated, the

Foreword

Although, the basic procedures for meat handling and processing are well established, the concerns today of meat quality and safety are uppermost in our minds. These can only be addressed through application of scientific techniques.

The book entitled, "*Practical Handbook of Meat Science and Technology*" authored *by J. Sahoo, D.K. Sharma and M.K. Chatli* of Department of Livestock Products Technology, College of Veterinary Science, Guru Angad Dev Veterinary and Animal Sciences University, Ludhiana, deals with information on composition and properties of meat and meat products and describes methods used for assessing their quality including issues of packaging. The compilation of this handbook is timely and strategic. The book shall be of great use to the teachers, researchers and students associated with meat science and technology.

V. K. Taneja

Preface

The primary object of meat foods preservation is to supply wholesome, safe and acceptable meat foods to the consumer. In order to ensure that the processed foods conform to certain specific standards, quality control measures are being implemented in the country. In India, the concept of quality standards was introduced by the Government through the Agricultural Produce (Grading and Marketing) Act (AGMARK) in 1937. Subsequently Meat Food Products order (MFPO), 1973 was promulgated in exercise of the powers conferred by section 3 of the Essential Commodities Act, 1955. The Prevention of Food Adulteration Act, 1954, was also passed by Parliament. The Indian Standards Institution was established in 1947 and thereafter, Bureau of India Standard was started in 1987 with the main objective of framing quality standards and ensuring their effective implementation.

In order to implement quality control measures, the methods followed for analysis by various agencies should be standardized. Analytical and quality testing of meat products composition is one part and guarantee of product quality is another aspect. In the food industry there is need for analysis of components in both raw and processed products. All sorts of analysis techniques are necessary in the development of meat food products and in controlling the food safety aspects. In view of this, it is strongly felt to make available the desired information concerned to above at one place in the form of this book entitled, *"Practical Handbook of Meat Science and Technology"*. The first author of the book Dr. Jhari Sahoo has been associated with industry, teaching, research, training and quality control in the field of meat and meat products technology for nearly 27 years, who has been aware of their problems and needs.

This is a handbook of methods of analysis and it offers the reader detailed descriptions of step by step procedures supported with cited references. This book contains 10 chapters. The chapter 1 describes about Common Laboratory Informations, while chapter 2 Proximate Composition, chapter 3 Physico-chemical Properties, chapter 4 Minerals, Vitamins and Enzymes, chapter 5 Food Additives and Preservatives, chapter 6 Biochemical Methods, chapter 7 Microbiological Quality, chapter 8 Molecular techniques in Meat Industry, chapter 9 about Sensory Evaluation Methods and at last Chapter 10 contains Testing of Packaging Films. This book can be used as a primary text book for B.V.Sc. & A.H., M.V.Sc. (LPT) and Ph.D. (LPT) students in their practical classes and research works of their projects. Besides, this book is of immense help to all the faculty members of all the Veterinary Colleges of the country who are engaged in the teaching/research/extension activities in the area of Livestock Products Technology/Animal Products Technology/Meat Science and Technology/Food Science and Technology.

I am thankful to both of my co-authors Dr. D.K. Sharma and Dr. M.K. Chatli for their sincere efforts and dedication in contributing, computing and editing the manuscripts of the book.

I am dedicating this book to my darling wife Pravasini, for her constant support and encouragement.

Dr. J. Sahoo

E-mail: sahoo_ j1203@yahoo.com; jsahoogadvasu@rediffmail.com

Dr. D.K. Sharma

Dr. M.K. Chatli

Contents

Chapter 1

Common Laboratory Informations

1. General Instructions

Some instructions for working in the laboratory are given here and the students are advised to read them carefully before they begin their work in the laboratory.

- ☆ Apparatus to be used should be neat and clean.
- ☆ For pipetting rubber bulb should be used.
- ☆ The small amount of liquid remaining in the pipette should not be blown out in standardizing the pipettes; allowance is always made for this. However, a frosted ring near the top of recently manufactured pipettes indicates that the last drop is to be blown out.

Table 1.1: Drainage Time of Various Pipettes

Capacity (ml)	2.0	10.0	20.0	25.0	50.0
Appx. Time of drainage (sec.)	10	20	28	30	35

- ☆ Most measuring devices are made of glass which has a small temperature coefficient, *e.g.* a soft glass vessel will lead to a change in volume by about 0.003 per cent per degree in temperature; with heat resistant glass the change is about one-third of this. As a general rule, the heating of caliberated equipment should be avoided. Rapid cooling can change the glass structure and cause changes in volume.

2. Laboratory Safety

1. Laboratory should be well ventilated and fitted with exhaust fans for effective removal of fumes.
2. Use apron and other devices like gloves, goggles etc. depending upon the material to be handled.
3. Do not add water to acids. Keep acids off skin and protect eyes from spattering. If acids are spilled on the skin, wash off immediately with tap water. Gaseous nitrogen oxides from HNO_3 can cause severe lung damage. Copious flow of fumes is there when both concentrated HNO_3 and HCl are mixed together.
4. If acid falls on clothes, neutralize the same with few drops of dilute ammonia solution or some other weak alkali solution.
5. If acid spills on the floor or the table, neutralize it with some weak alkali and wipe off with duster.
6. If you happen to suck acid into your mouth during pipetting, wash your mouth quickly with water and then rinse with a weak solution of washing soda.
7. Use fume hood to protect against any type of fumes.
8. Avoid use of equipments for purposes other than intended.

3. Cleaning Laboratory Wares

The glass and porcelain wares should be thoroughly washed with some detergent then these should be extensively rinsed with tap water followed by further rinsing with distilled water.

If a grease film remains after cleaning with detergent, a cleaning solution consisting of sodium or potassium dichromate in concentrated sulphuric acid may be used. After this rinsing is necessary in order to remove the last traces of dichromate ions, which adhere strongly to glass or porcelain surface.

Preparation of Cleaning Solution

Dissolve 40 g of sodium or potassium dichromate (commercial grade) in 150 ml of water (with a little heat if necessary) cool to room temperature and add slowly 230 ml of conc. H_2SO_4 (CP).

Precaution

This solution always be prepared over a sink.

Note

1. Cleaning solution should be discharged when it acquired green color of chromium ion.
2. Cleaning solution is most effective when warmed up to about 70°C. At this temperature it rapidly attacks plant and animal matter and thus it is potentially a dangerous preparation.
3. Spillage, if occurs, should be cleaned with water.

4. Laboratory Instructions

1. Remember that you are in a microbiological laboratory and are working with microbes, some of which may be capable of producing disease. Therefore, take extreme care in handling the cultures, slides and other materials, which you may use or come in contact in the laboratory.
2. Hand over all used slides and cover glasses to the attendant or discard them in the specific container.
3. Avoid spillage of materials or stains.
4. Wash your hands before you come and after you leave the laboratory.
5. In case of an accident, please bring it to the attention of the instructor.
6. Take extreme care while you handle the Microscope and its eye pieces.
7. If you have used an oil immersion lens make sure that you have cleaned the lens with xylene before you return the instrument.

8. Always make it a rule to flame the inoculating needle before and after use. Also, always open a culture tube or flask near the flame.
9. The instruction book is provided so that you can prepare and plan your laboratory work. Therefore, read the instructions carefully before you come to the laboratory. Your laboratory report on the exercise you perform should be submitted for evaluation before performing the succeeding exercise.
10. If you can afford, use a white laboratory coat/apron and a separate piece of cloth to clean up the workbench.

5. Preparations of Normal Solutions

Table 1.2: Strength of Aqueous Solution of Common Acids and of Ammonium Hydroxide

Substance	*Specific Gravity*	*Normality*	*Percent W/W*	*Millililtres to be Taken to Make One Litre of N (approx.) Solution*
Hydrochloric acid	1.18	11.3	35	89
	1.16	–	32	98.2
Nitric acid	1.42	16.0	70	63
Sulphuric acid	1.84	36.0	96	28
Phosphoric acid	1.69	41.1	85	23
Acetic acid	1.05	17.4	99.5	58
Ammonium hydroxide	0.90	14.3	27	71

(*i*) Preparation of a Decinormal (N/10) Hydrochloric Acid (HCl) Solution

Principle

Since Hydrochloric acid (HCl) is not a primary standard substance it has to be standardized against a primary standard. Here Na_2CO_3 is taken as primary standard. HCl reacts with Na_2CO_3 in following manner.

$$Na_2CO_3 + 2\,HCl \longrightarrow 2\,NaCl + H_2O + CO_2$$

Reagents and Materials

1. Concentrated HCl AR*
2. Methyl red or methyl orange indicator
3. N/10 Na_2CO_3
4. Measuring flask
5. Burette
6. 10 ml pipette
7. Conical flask

*Concentrated HCl assay is 35.4 per cent Specific gravity is 1.176

$$V = 100(MN)/bPd = \frac{100 \times 36.46 \times 0.1}{1 \times 35.4 \times 1.176} \quad 8.78 \text{ ml HCl}$$

Procedure

- ☆ 4.5 ml of concentrated Hydrochloric acid (HCl) AR was pippetted out to a measuring flask of 500 ml capacity in which some water was taken previously.
- ☆ Precaution is always taken to add acid into water, never the reverse.
- ☆ Volume was made upto ½ litre.
- ☆ Then in a rinsed and dried burette, this HCl solution was taken.
- ☆ In a conical flask, exactly 10 ml Na_2CO_3 N/10 solution was taken and 2 drops of methyl red indicator was added.
- ☆ This standard solution of Na_2 CO_3 was titrated against the Hydrochloric acid (HCl) solution.
- ☆ End point of the titration reached when the titrated solution became orange. Read color.
- ☆ Then from the known volume of HCl consumed, the strength was calculated by the formula $N_1V_1 = N_2V_2$.
- ☆ The strength of HCl was found to be N/8.2
- ☆ Then to make it N/10, the volume of HCl required is calculated by the formula $N_1V_1 = N_2V_2$.
- ☆ The desired volume was made with distilled water. The solution was labeled as "N/10 HCl".

Result

It is observed that volume of HCl consumed is 8.2 ml

Calculations

$N_1V_1 = N_2V_2$

$N/10 \times 10ml = N_2 \times 8.2\ ml$

Or $N_2 = N/10 \times 10ml/8.2ml = N/8.2$

Strength of HCl was found to be N/8.2

To prepare 250 ml of N/10 HCl solution

The Vol. of N/8.2 HCl required = $V_2 = N_1V1/N_2 = N/10x250/N/8.2$

$N/10 \times 250 \times 8.2/N = 25 \times 8.2 = 205$ ml

So, 205 ml N/8.2 HCl + 45 ml distilled water = 250 ml N/10HCl

(*ii*) Preparation of Decinormal Sulphuric Acid (N/10 H_2SO_4) solution

Principle

H_2SO_4 is not a primary standard and is unstable. It is to be standardized against Na_2CO_3

Reaction

$$H_2SO_4 + Na_2CO_3 \longrightarrow Na_2SO_4 + CO_2 + H_2O$$

Regents and Materials

1. Na_2CO_3 (N/10)
2. Mixed indicator (Bromocresol green and Methyl red 2:1)
3. H_2SO_4
4. Volumetric flask (500 ml)
5. 100 ml conical flask
6. Burette and burette stand
7. Pipette
8. Funnel

Procedure

- ✰ In volumetric flask having some distilled water 1.4 ml concentrated H_2SO_4 was pippetted and volume was made up to 500 ml with distilled water.
- ✰ It was standardized by titrating against N/10 Na_2CO_3 using mixed indicator.
- ✰ Normality of H_2SO_4 was calculated and the amount of water to be added was also calculated.
- ✰ It was again titrated against N/10 Na_2CO_3 to recheck the normality.

Inference

Since 10 ml H_2SO_4 was required to titrate against 10 ml of N/10 Na_2CO_3 normality of H_2SO_4 was N/10.

(*iii*) Preparation of a N/70 Sulphuric Acid (H_2SO_4) Solution

Reagents and Materials

1. Analytical balance
2. Clean dry beaker
3. 250 ml volumetric flask
4. Burette and pipette
5. Sodium carbonate
6. Sulphuric acid
7. Methyl orange indicator

(*a*) Approximately N/70 H_2SO_4 Solution

- ✰ Molecular weight of H_2SO_4 =2.016 + 32.006 + 64.00 = 98.082
- ✰ Gram equivalent = 98.082/2 = 49.041

Therefore one litre of 'N' solution will contain 49.041g of H_2SO_4

- ✰ N/70 solution will contain 49.041/70 = 0.7005g of H_2SO_4
- ✰ Specific gravity of H_2SO_4 = 1.84
- ✰ Expressing in terms of volume = 0.7005/1.84
- ✰ Purity of acid = 95 per cent

- Therefore, $N/70\ H_2SO_4 = \dfrac{0.7005 \times 100}{1.84 \times 95} = 0.4$ ml
- Therefore for preparing one litre approximately solution of $N/70\ H_2SO_4$ roughly 0.4 ml concentrated H_2SO_4 AR is required.

Procedure

- A clean dry one litre volumetric flask is half filled with distilled water.
- Measures exactly 0.4 ml concentrated H_2SO_4 AR in a graduated cylinder and gradually add into the flask avoiding any spurting.
- Allow to cool for few min.
- Make up the volume up to the mark with distilled water and allow to stand overnight after thorough mixing.
- This is ready for titration with primary standard.

Primary Standard

N/70 Na_2CO_3

- Molecular weight of Na_2CO_3 = 45.982 + 12.011 + 48.00 = 105.993g
- Gram equivalent weight = 105.993/2 = 52.99g say 53g

Therefore one litre of 'N' solution will contain 53.00g of Na_2CO_3

- 250 ml will contain53.00/4g of Na_2CO_3
- Therefore N/70 solution (250 ml) = 55/4×70 = 0.1893g
- Weigh 0.1893g of Na_2CO_3 in 250 ml volumetric flask and make the volume up to the mark.

Standardization of N/70 H_2SO_4

- Accurately take 100 ml of approximately $N/70\ H_2SO_4$ in a clean dry beaker.
- Rinse the burette with this solution and fill the burette with solution.
- Pipette out exactly 10 ml primary standard (N/70 Na_2CO_3) in a clear dry conical flask and add 1-2 drops of methyl orange indicator into the conical flask.

- ☆ Note the initial reading and titrate with primary standard solution till the end point (yellow color) appears.
- ☆ Note the volume of approximately N/10 H_2SO_4 consumed to equalize the primary standard taken from the titration.
- ☆ Repeat the titration 3-4 times till a constant reading is obtained.
- ☆ Calculate from titer reading the amount of distilled water required for diluting the approximately N/70 H_2SO_4 solutions and then standardize.

(*iv*) 1N Oxalic Acid (63.023g $H_2C_2O_4 \cdot 2H_2O$ per litre)

Decinormal or less-concentrated solution are unstable and should be prepared fresh when needed. More concentrated solutions may deposit some of the acid when cooled to low temperature, but they are fairly stable at room temperature when protected from light.

(*v*) Preparation of Decinormal Sodium Hydroxide (N/10 NaOH) Solution

Principle

Since NaOH is not a primary standard; it is not in a pure form, hence need standardization. N/10 HCl is used to standardize NaOH.

Reaction

$$NaOH + HCl \longrightarrow NaCl + H_2O$$

Reagents and Materials

1. N/10 HCl
2. NaOH
3. Phenolphthalein indicator (0.05g in 100ml ethanol)
4. 500 ml volumetric flask
5. Burette and Burette stand
6. 10 ml pipette
7. 100 ml conical flask
8. Glass rod.

Procedure

- ☆ Two grams of NaOH was weighed, then it was transferred to volumetric flask and some distilled water was added to dissolve it.

- ✰ The volume was made to 500ml with distilled water.
- ✰ Approximately 50 ml burette was filled with this NaOH.
- ✰ It was titrated against 10 ml of N/10 HCl after adding 1-2 drops of phenolphthalein indicator in a conical flask.
- ✰ End point was taken as change of color from colorless to pink.
- ✰ Three readings were taken and mean was calculated.
- ✰ From the amount of NaOH used, its normality was calculated.
- ✰ If required water was added till the desired normality was obtained.

Inference

Since 10 ml of NaOH was required to titrate 10 ml of N/10 HCl; the normality of NaOH is N/10.

(*vi*) Preparation of a Decinormal Sodium Carbonate (N/10 Na_2CO_3) Solution

Theory

$$Na_2CO_3 + 2\,HCl \longrightarrow 2\,NaCl + H_2O + CO_2$$

Formal weight or Molecular weight of $Na_2CO_3 = 2 \times 23 + 12 + (16 \times 3)\ 48 = 106$

Basicity (number of replaceable H_2) = 2

Equivalent weight (Eq.wt.) of $Na_2CO_3 = 106/2 = 53$

So, 53 g of Na_2CO_3 dissolved in 1 litre of distilled water gives 1N solution. For preparation of N/10 solution, 5.3 g of Na_2CO_3 was taken. Since Na_2CO_3 is hygroscopic it should be kept at 200°C for 1h.

Reagents and Materials

1. Na_2CO_3 Analytical reagent (AR)
2. Beaker 100ml
3. Measuring flask 1000 ml
4. Analytical balance

Procedure

- ✰ About 6 g of Sodium carbonate (Na_2CO_3) was weighed in a balance. Then taken in a beaker and was kept at 200°C for 1 h.

- ☆ Then exactly 5.3 g of Na_2CO_3 was weighed and transferred to a 1 litre measuring flask.
- ☆ Then some water was poured into the flask till Na_2CO_3 was dissolved and volume was made up to 1 litre.
- ☆ At last the standard solution thus prepared was labeled as N/10 $Na_2 CO_3$ solution.

(*vii*) 0.1 N Potassium Dichromate (4.9037 g of $K_2 Cr_2O_7$ per litre)

Dry crystals of potassium dichromate at 120 to 140°C for 2 to 4 hr. Cool in a desiccators and weight 5 g to a nearest milligrams. Dissolve in about 200 ml of water, transfer to a 1-litre volumetric flask, dilute to volume and mix thoroughly.

$$\text{Normally} = \frac{\text{Wt. of Potassium Dichromate}}{49.037}$$

(*viii*) Preparation of Decinormal Potassium Permagnate (0.1N $KMnO_4$) Solution

Principle

It is based on oxidation-reduction. $KMnO_4$ is an oxidizing agent, which is allowed to react with reducing agent $Na_2C_2O_4$. When whole $Na_2C_2O_4$ is oxidized the purple color of $KMnO_4$ shows the end point. Presence of H_2SO_4 is must

$$2\,KMnO_4 + 3\,H_2SO_4 \longrightarrow K_2SO_4 + 2MnSO_4 + 3H_2O + 5(O)$$

$$Na_2C_2O_4 + H_2SO_4 \longrightarrow H_2C_2O_4 + Na_2SO_4$$

$$H_2C_2O_4 + O \longrightarrow H_2O + CO_2$$

$$2KMnO_4 + 5Na_2C_2O_4 + 8H_2SO_4 \longrightarrow K_2SO_4 + 2MnSO_4 + 5Na_2SO_4 + 8H_2O + 10\,CO_2$$

It means $2KMnO_4 = 5Na_2C_2O_4$

$$\frac{KMnO_4}{5} = \frac{Na_2C_2O_4}{2}$$

Equivalent weight of $KMnO_4$ is 1/5 molecular weight = 32

Equivalent weight of $Na_2C_2O_4$ is 1/2 molecular weight = 69

Reagents

1. *0.1N $KMnO_4$*: Weigh 0.85 g of solid $KMnO_4$. Transfer it to beaker containing 250ml-distilled water. Then heat up to boiling and cool to room temperature. Filter the solution through glass wool and keep in dark brown bottle.
2. *Sodium oxalate*: Analytical reagent sodium oxalate crystals are kept in oven for 1h. Transfer it to dessicator for cooling. Weigh 0.67g, dissolve in water and make volume 100ml.
3. *Sulphuric acid* (1:1): Slowly add 50ml of concentrated Sulphuric acid to 50 ml of distilled water in a beaker. Then transfer it to a reagent bottle.

Procedure

- Burette is rinsed thoroughly with $KMnO_4$.
- Fill the burette with 0.1N $KMnO_4$. Take 10 ml of $Na_2C_2O_4$ solution in a conical flask and add 10ml of 1:1 H_2SO_4.
- Warm to boiling.
- First add 2ml slowly for getting speedy reaction thereafter add rapidly till end point is obtained (faint pink color).

Table 1.3: Observations

Sl.No	Initial Burette Reading	Final Burette Reading	ml of $Na_2C_2O_4$ Taken	ml of $KMnO_4$ Used
1.				
2.				
3.				

Calculations

Volume of $KMnO_4$ used = V_1

Normality of $KMnO_4$ = N_1

Volume of $Na_2C_2O_4$ = V_2

Normality of $Na_2C_2O_4$ = N_2

$$N_1V_1 = N_2V_2$$

$$N_1 = \frac{N_2V_2}{V_1} = \text{say x}$$

To make exactly 0.1N $KMnO_4$ $\quad N_1V_1 = N_2V_2$

$$0.1 \times 250 = x \times V_2$$

$$V_2 = \frac{250 \times 0.1}{x} = \text{say y ml}$$

So 250–y ml of distilled water is added to y ml of approximate $KMnO_4$ to get exact 0.1N $KMnO_4$

Precautions

- ☆ Since $KMnO_4$ solids contain impurities it is necessary to boil and filter.
- ☆ $Na_2C_2O_4$ must be dried
- ☆ Before titration, solution must be warmed but not boiled.
- ☆ Pink color sometimes is temporary so at least it should remain for 30 sec to ensure correct end point.

(*ix*) Preparation of N/40 Potassium Permanganate (N/40 $KMnO_4$) Solution

Principle

$KMnO_4$ is not a primary standard and is unstable. It is to be standardized against Sodium oxalate (N/40)

Reactions

$$2\,KMnO_4 + 3H_2SO_4 \longrightarrow K_2SO_4 + 2MnSO_4 + 3H_2O + 5(O)$$

$$(COO\,Na)_2 + (O) + H_2SO_4 \longrightarrow 2CO_2 + H_2O + Na_2SO_4$$

Regents and Materials

1. Sodium Oxalate
2. $KMnO_{4,}$
3. 5 per cent H_2SO_4
4. Volumetric flask (500 ml)
5. 100 ml conical flask

6. Burette and Burette stand
7. Pipette
8. Beaker

Procedure

- For making the 5 per cent H_2SO_4 solution we took 10 ml of 95 per cent H_2SO_4 solution and made it to volume 200 ml with distilled water.
- Then weighed 0.8375 g of Sodium oxalate and took it in a volumetric flask, dissolved it in small amount of water and made it to a volume of 500 ml.
- Took 10 ml of Sodium oxalate and 10 ml of 5 per cent H_2SO_4 and heat it on the Soxhlet apparatus hot plate and fill the 50 ml burette with $KMnO_4$ solution.
- Then titrate it against the Sodium oxalate + 5 per cent H_2SO_4 solution.
- Took 3 readings and calculate the mean.
- From the used $KMnO_4$ solution, calculate the normality and from the left over amount of $KMnO_4$ solution calculate the normality.
- Further amount of water is added to get the desired normality.
- Again titrate it against the sodium oxalate and find out the normality again.
- If required again add small amount of water and get the desired N.

Inference

Since 10 ml of $KMnO_4$ was required to titrate against 10 ml of N/40 Sodium oxalate so normality of $KMnO_4$ was N/40.

(*x*) Preparation of Decinormal Sodium Thiosulphate (N/10 $Na_2S_2O_3$) Solution

Principle

Standardization of sodium thiosulphate is done by using $K_2Cr_2O_7$ of known strength. HCl and KI produce HI. HI is oxidized by $K_2Cr_2O_7$ and gives iodine. I_2 reacts with $Na_2S_2O_3$ to give NaI. At end point, blue to colorless is obtained when all the $Na_2S_2O_3$ has been used.

Reactions

$$K_2Cr_2O_7 + HCl \longrightarrow KCl + 2CrCl_3 + 4H_2O + 3(O)$$

$$KI + HCL \longrightarrow KCl + HI \times 6$$

$$2HI + O \longrightarrow H_2O + I_2 \times 3$$

$$Na_2S_2O_3 + I_2 \longrightarrow Na_2S_4O_6 + 2NaI \times 3$$

$$K_2Cr_2O_7 + 14\,HCl + 6\,KI + 6\,Na_2S_2O_3 \longrightarrow 8\,KCl + 2\,CrCl_3 + 7\,H_2O + 6\,NaI + 3\,Na_2S_4O_6$$

$$Cr_2O_7^{-2} + 14H^+ + 6\,I \longrightarrow 2Cr^{+3} + 3I_2 + 7H_2O$$

Reagents

1. *Approximate 0.1N solution of $Na_2S_2O_3$*: 6.25g of sodium thiosulphate were dissolved in 250ml water. It was boiled and solution was stored in dark reagent bottle in dark.
2. *Potassium dichromate*: It was dried in the oven for 60 min. 0.49g was accurately weighed and placed in measuring flask and volume was made up to 100 ml.
3. *Starch*: One gram starch was mixed with cold water. It was kept in 100 ml boiling water with constant stirring.
4. *Potassium iodide*
5. *Hydrochloric acid*: 25 ml of concentrated HCl was added slowly to 250 ml of water with constant stirring

Procedure

- ✰ 10 ml of $K_2Cr_2O_7$ was taken in a conical flask.
- ✰ One gram of KI was added. 20 ml of 1N HCl was added and kept in the dark for 10 min.
- ✰ Burette was rinsed and filled with sodium thiosulphate solution.
- ✰ After 10 min, solution was titrated and starch solution was added after the most of iodine has been consumed.
- ✰ Color change is blue to colorless. Three readings were observed to get the concurrent reading.

Table 1.4: Observations

Sl.No	ml of $K_2Cr_2O_7$	Initial Burette Reading	Final Burette Reading	ml of $Na_2S_2O_3$ Used
1.				
2.				
3.				

Calculations

$$N_1V_1 = N_2V_2$$

$N_1 = 0.1$
$V_1 = 10$ ml
$N_2 = ?$
$V_2 = x$ ml

$$N_2 = \frac{10 \times 0.1}{x \text{ ml}}$$

Normality of $Na_2S_2O_3$ =

Precautions

- ✰ Starch solution must be added after most iodine has been consumed
- ✰ It must be kept in dark for at least 10 min to complete reaction

(*xi*) Preparation of Decinormal Silver Nitrate (0.1N $AgNO_3$) Solution

Principle

It is a precipitate titration. Potassium chromate is used as indicator. At the end point, chromate ions combines with silver ion forming reddish brown silver chromate precipitates. $AgNO_3$ first reacts with KCl to give AgCl precipitate. As soon as KCl is exhausted, Potassium chromate reacts with $AgNO_3$ forming reddish brown precipitate of silver chromate

Reactions

$$AgNO_3 + KCl \longrightarrow AgCl + KNO_3$$

$$2AgNO_3 + K_2CrO_4 \longrightarrow 2KNO_3 + Ag_2CrO_4$$

Reagents

1. *Approximate 0.1 N $AgNO_3$ solution*: 1.7g of silver nitrate were dissolved in 50 ml of distilled water. Volume was made up to 100ml.
2. *Potassium chloride*: Dry AR Potassium chloride was kept in oven for 1h. 0.7g was accurately weighed and dissolved in water.
3. *Potassium chromate solution*: 2.5g of Potassium chromate was weighed. It was dissolved in water and volume was made up to 50ml.

Procedure

- ☆ Silver nitrate solution was used for rinsing the burette.
- ☆ It was then filled in the burette.
- ☆ 10ml of KCl was taken in a flask and 1ml of potassium chromate was added.
- ☆ At end point of titration, red brown precipitate appears. Blank was also run in the same way and reading of burette was noted.

Table 1.5: Observations

Blank value =

Sl.No	*ml of KCl Used*	*Initial Burette Reading*	*Final Burette Reading*	*ml of $AgNO_3$ Used*
1.				
2.				
3.				

Calculations

$$N_1V_1 = N_2V_2$$

N_1 = ?

V_1 = x ml–blank reading = say y

$N_2 = 0.1$

$V_2 = 10$

$$N_1 = \frac{0.1 \times 10}{y}$$

Normality of $AgNO_3$ =

Precautions

- ✰ KCl must be weighed accurately in order to make correct 0.1N solution
- ✰ End point must be determined correctly
- ✰ $AgNO_3$ must be weighed slightly more to have 0.1 or more normality. If normality is more, it may be reduced by adding water. If normality is less there is no remedy for it.

Alternate Methods of Preparation of 0.1N Silver Nitrate Solution

Mohr Method

Dissolve slightly more than the required quantity (17.2 g instead of 16.989 g) of reagent grade silver nitrate in water and dilute to 1000 ml in a volumetric flask. Weight accurately 0.25 g of NaCl (dried at 110°C before weighing) and transfer to a 250-ml conical flask with 50 ml of a halogen-free water. Add 1 ml of 5 per cent solution of potassium chromate (K_2CrO_4) in water as indicator. Titrate with the silver nitrate solution until a perceptible pale red-brown color appears. Carry out a blank titration using 75-100 ml of distilled water and 1 ml of indicator. Calculate the normality from the net volume.

$$\text{Normality} = \frac{\text{Wt of NaCl (g)} \times 1000}{\text{ml of } AgNO_3 \times 58.45}$$

Volhard Method

Prepare 0.1 N Standard solution of potassium of ammonnium thiocyanate by dissolving 7.613 g of ammonium thiocyanate (NH_4 CNS) OR 9.719 G of potassium thiocyanate (KCNS) in 1000 ml of water. Pipette 50 ml of standard 0.1 N silver nitrate solution (17.2 g/ litre), add 2 ml of ferric alum [$FeNH_4$ $(SO_4)_2$. 12 H_2O] indicator (saturated solution in water) and 5 ml of HNO_3 (1+1), and titrate with the thiocyanate solution until the solution shows a pale rose color after vigorous shaking. This gives the direct titre (A).

To standardize, weight accurately 0.125 g of NaCI (dricd at 110°C before weighing) and transfer to a 250-ml conical flask with 50 ml of water. Add 5 ml of HNO_3 (1+1) and 50 ml of 0.1 N silver nitrate solution used before. Mix and allow it to stand for a few minutes in the dark. Filter through gooch crucible prepared with asbestos and previously rinsed with 2 per cent HNO_3. Wash repeatedly with2 per cent HNO_3 until the filtrate is about 150ml. Titrate the residual silver nitrate with the thiocyanate solution using 2 ml of ferric alum as indicator (B). Find out the net volume of silver nitrate reacted with sodium chloride by subtracting B from A and calculate the normality of silver nitrate.

$$\text{Normality of } AgNO_3 = \frac{\text{Wt. of NaCl (g)} \times 1000}{\text{ml of } AgNO_3 \times 58.45}$$

(*xii*) Preparation of Decinormal Iodine (0.1N I_2) Solution

Principle

If strong oxidizing reagent is treated in neutral or acid medium with excess of iodide ion, the later reacts as a reducing agent and the oxidant will be quantitatively reduced. Equivalent amount of iodine is liberated, and then titrated with standard solution of a reducing agent ($Na_2S_2O_3$) At end point, colorless solution is obtained by using starch as indicator.

Reaction

$$I_2 + 2\,Na_2S_2O_3 \longrightarrow Na_2S_4O_6 + 2\,NaI$$

Reagents

1. Potassium Iodide
2. Iodine
3. $Na_2S_2O_3$ solution

Procedure

- ☆ 2g Potassium Iodide was weighed. It was dissolved in water kept in beaker. 1.3 g Iodine was mixed in it and was stirred to dissolve the iodine. Volume was made to 100ml.
- ☆ Burette was rinsed by $Na_2S_2O_3$ solution and filled up.
- ☆ Starch was added as an indicator. 10 ml of 1N HCl was added. It was titrated and reading was noted.

Table 1.6: Observations

Sl.No	ml. of Iodine of Solution taken	Initial Burette Reading	Final Burette Reading	ml of $Na_2S_2O_3$ Used
1.				
2.				
3.				

Calculations

$$N_1V_1 = N_2V_2$$

$N_1 = ?$
$V_1 = 10$ ml
$N_2 = 0.151$
$V_2 = x$ ml

$$N_1 = \frac{0.151 \times x\ \text{ml}}{10}$$

Normality of Iodine solution = ?

Precautions

- ☆ Solution must be kept in dark, away from light
- ☆ First KI must be dissolved in the water then I_2 crystals must be added in order to have better solubility

Indicators

Preparation of common indicators and their pH range are given in Tables 1.7 and 1.8 respectively.

Table 1.7: Preparation of Common Indicators

Sl.No.	*Indicator*	*pH Range*	*Preparation*
1	Phenolphthalein	8.3-10.0	Dissolve 1g in 100 ml ethanol
2	Methyl red	4.4-6.0	Dissolve 1g in 100 ml of 95 per cent ethyl alcohol. This indicator is easily reduced with loss of color and titration should be carried out immediately after it is added to the solution.
3	Methyl orange	2.9-4.0	Dissolve 0.5g in 1000 ml of water
4	Bromophenol blue	3.0-4.6	Dissolve 0.1g in 25 ml of water and dilute to 100 ml with water
5	Mixed methyl red-methylene blue indicator		Mix equal volumes of aqueous 0.2 per cent methyl red and aqueous 0.1 per cent methylene blue.
6	Methylene blue		Make up 1g of pure material to 100 ml with distilled water.
7	Starch solution (0.5 per cent)		Mix 0.5 g of soluble starch with about 15ml of cold water and pour into 100 ml of hot water. Boil for 1 to 2 min.

Table 1.8: pH Range of Common Indicators

Name	*pH Range*	*Color in Acid Solution*	*Color in Alkaline Solution*
Acid cresol red	0.2-1.8	Red	Yellow
Acid meta cresol purple	1.2-2.8	Red	Yellow
Thymol blue	1.2-2.8	Red	Yellow
Benzo yellow	2.4-4.0	Red	Yellow
Methyl orange	2.9-4.6	Red	Orange/yellow
Bromocresol green	3.8-5.4	Yellow	Blue
Bromophenol blue	3.0-4.6	Yellow	Blue
Methyl red	4.4-6.0	Red	Yellow
Chlorophenol red	5.2-6.8	Yellow	Red
Bromocresol purple	5.2-6.8	Yellow	Purple
Bromothymol blue	6.0-7.6	Yellow	Blue
Phenol red	6.8-8.4	Yellow	Red
Cresol red	7.2-8.8	Yellow	Red

Contd...

Table 1.8–Contd...

Name	*pH Range*	*Color in Acid Solution*	*Color in Alkaline Solution*
Metacresol purple	7.6-9.2	Yellow	Purple
Thymol blue	8.0-9.6	Yellow	Blue
Phthalein red	8.6-10.2	Yellow	Red
Tolyl red	8.6-11.6	Red	Yellow
Parazo orange	11.0-12.6	Yellow	Orange
Acyl blue	12.0-13.6	Red	Blue

Chapter 2

Proximate Composition

1. Moisture Estimation

Methods of Moisture Determination

Numerous reviews of moisture analysis methods have been published in Handbook of Moisture determination and Control, in four volumes. Detailed methodologies for specific food products can be found in the Official Methods of Analysis of AOAC International.

Table 2.1: Classification of Analytical Methods for Moisture Determination

	Classification by Four Major Principles		*Classification by Direct/ Indirect Procedures*
1.	Drying methods	1.	Direct methods
	oven drying		gravimetric methods
	vacuum drying		oven drying
	freeze-drying (lyophilization)		air oven
	chemical desiccation		vacuum oven
	thermogravimetric analysis		freeze-drying
2.	Distillation methods		thermogravimetric analysis
	direct distillation		chemical desiccation
	reflux distillation		distillation methods

Contd...

Table 2.1–Contd...

	Classification by Four Major Principles		*Classification by Direct/ Indirect Procedures*
3.	Chemical methods		direct distillation
	Karl Fischer titration		reflux distillation
	penetration of acetylene		chemical titration method
4.	Physical methods		Karl Fischer
	infrared absorption		extraction method
	Near-infrared reflectance		gas chromatography
	gas chromatography	2.	Indirect methods
	nuclear magnetic resonance		spectroscopic methods
	refractometry		infrared absorption
	neutron scattering		near-infrared reflectance
	electrical		nuclear magnetic resonance
	microwave absorption		mass spectrometry
	dielectric capacitance		electrical methods
	conductivity		microwave absorption
	cryoscopic methods		conductivity
			dielectric capacitance
			sonic and ultrasonic methods
			neutron scattering
			refractometry
			cryoscopic methods

Analytical methods of moisture determination can be classified in two ways as shown in Table 2.1. One of these ways is by the four major analytical principles: drying, distillation, chemical, and physical methods. The other is by direct and indirect procedures based on the underlying scientific thereby. The direct methods for moisture analysis normally involve removing water from the solid food samples by drying, distillation, extraction, or other method, and its quantity is measured by weighing, titration, and so forth. For the indirect methods, moisture is not removed from the sample and quantified directly, but rather the properties of the food that depend on either the amount of water or number of hydrogen atoms are measured. These indirect methods must be calibrated against standard moisture values that have been precisely determined using one or more of the direct methods. Therefore, the accuracy of indirect methods is dependent on the values of direct measurements against which they are calibrated.

Direct methods usually yield accurate and even absolute values for moisture determination, but they are mostly manual and time consuming. On the other hand, indirect methods are rapid, nondestructive, and offer the possibility of automation for continuous determination. For the convenience of presentation, methods of moisture determination are classified here into two categories, namely direct and indirect. Advantages and disadvantages of each individual method under the two classification are described in Tables 2.2 and 2.5 .

(*i*) Direct Methods

(*a*) Air-Oven Drying

Owing to its convenience air-oven drying methods are one of the most common and widely used methods for routine moisture determination in laboratories around the world. Drying can be accomplished using either convection type ovens or forced-draft ovens. The ovens should be thermally regulated to ±0.5°C and have minimal temperature variations (<±3°C) within the oven. Forced-draft ovens offer a more consistent temperature throughout the oven than that of convection ovens. Modern drying ovens are usually heated by electricity or infrared heaters and can be equipped with built-in balances for routine and fast analysis, assuming that the food is stable. Because the principle of oven drying is based on weight loss, the sample needs to be thermally stable and should not contain significant amount of volatile compounds.

The general operational procedures for the conventional method of moisture determination using a drying oven and analytical balance involve sample preparation, weighing, drying, cooling, and reweighing.

(*b*) Vacuum-Oven Drying

Vacuum-oven drying is generally the standard and most accurate drying method for moisture analysis of foods. Many of the drawbacks associated with air-oven drying can be overcome by this method. Typically, vacuum drying heats foods at 98-102°C and 25-100 mm Hg pressure. Lower temperatures (60-70°C) can be used for products moisture pressure evaporates quicker and drying times can be dramatically reduced. Even though it is impossible to obtain an absolute moisture level by drying methods, vacuum drying can

Table 2.2: Advantages and Disadvantages of Direct Methods for Moisture Determination

Method	*Advantgaes*	*Disadvantages*
A. Oven drying	1. Standard conventional method	1. Variation of temperature due to particle size, sample weight, position in oven, etc.
	2. Convenient	2. Difficult to remove all water.
	3. Relative speed and precision	3. Loss of volatile substances during drying
	4. Accommodates large number of samples	4. Decomposition of sample (*e.g.* sugar)
	5. Attains desired temperature more rapidly	
B. Vacuum-oven drying	1. Lower heating temperatures possible	1. Possible volatile loss
	2. Prevents sample decomposition	2. Lower number of samples than for oven drying
	3. Uniform heating and constant evaporation	3. Drying efficiency reduced for high-moisture foods
C. Freeze-drying	1. Excellent for sensitive, high-value liquid foods	1. Expensive
	2. Preserves texture and appearance	2. Long drying time
	3. No foaming	3. Sample must be initially frozen
	4. No case hardening	4. Most applicable to high-moisture foods
	5. No oxidation	
	6. No bacterial changes during drying.	

Contd...

Table 2.2–Contd...

Method	Advantgaes	Disadvantages
D. Distillation methods	1. Determines water directly rather than weight loss	1. Low precision of measuring device.
	2. Apparatus is simple to handle	2. Organic solvents, such as toluene, pose a fire hazard.
	3. Accuracy may be greater than that of oven drying	3. Organic solvents may be toxic.
	4. Takes relatively short time (30 min to 1 h) to determine	4. Can have higher results due to distillation of water-soluble components (*e.g.* glycerol and alcohol)
	5. Prevents oxidation of sample	5. Water droplets may adhere to internal surface of apparatus causing erroneous results.
	6. Not affected by environmental humidity	6. Emulsion may form.
	7. Suitable for samples containing volatiles	
E. Karl Fischer method	1. A standard method for moisture analysis	1. Chemicals of highest purity must be used for preparing the reagent
	2. Accuracy and precision are higher than for other methods	2. Titration end point may be difficult to determine visually
	3. Useful for determining water in fats and oils by preventing oxidation	3. Reagents is unstable and needs standardization before use.

Contd...

Table 2.2–Contd...

Method	*Advantgaes*	*Disadvantages*
	4. Once apparatus is set up, determination takes a few minutes.	4. Titration apparatus must be protected from atmospheric moisture due to extreme sensitivity of reagent to moisture.
	5. Automated equipment available	5. Ascorbic acid and other carbonyls can react with reagents causing overestimation of moisture content.
F. Chemical desiccation	1. Can serve as reference standard for other methods	1. Requires long time to achieve constant dry weight
	2. Can be done at room temperature	2. Moisture equilibrium depends on strength of desiccant.
	3. Good for measuring moisture in substances containing volatile compounds.	
G. Thermogravimetric analysis	1. More automated than standard oven drying	1. Excellent for research, but not practical
	2. Weighing error is minimal because sample is not removed from oven.	2. Small sample may not be representative
	3. Sample size is small	3. Sample may decompose or oxidize
H. Gas chromatography	1. Analysis is rapid (takes 5-10 min per sample)	1. Unit cost per sample may be higher than for oven drying
	2. Results similar to those of conventional methods	2. Sample extraction required
		3. Requires expensive equipment

Table 2.3: Advantages and Disadvantages of Indirect Methods for Moisture Determination

Method	*Advantgaes*	*Disadvantages*
A. Refractometry	1. Determination takes only 5-10 min. (rapid)	1. Temperature sensitive
	2. Does not requires complex or expensive instrumentation	2. Requires uniformity of fluid samples
	3. Simple method	3. Solid samples (*e.g.*, meat) require homogenization in anhydrous solvent
	4. Reasonable accuracy	
	5. Excellent for high-sugar products	
B. Infrared absorption	1. Can perform multi- component analysis	1. Accuracy depends on calibration against reference standard
	2. Most versatile and selective	2. Temperature dependent
	3. Non-destructive analysis	3. Dependent on homogenization efficiency of sample
		4. Absorption band of water is not specific
C. Near-infrared reflectance	1. Rapid	1. Reflectance data is affected by sample particle size, shape, packing density, and homogeneity
	2. Precise	2. Interference between chemical groups (*e.g.*, hydroxyl and amine)
	3. Non-destructive	3. Temperature dependent

Contd...

Table 2.3–Contd...

Method	*Advantgaes*	*Disadvantages*
	4. No extraction required	4. Accuracy depends on calibration of standard samples
	5. Minimum sample preparation	5. Equipment is expensive
D. Microwave absorption	1. Non-destructive	1. Possible leakage of microwave energy during measurement
	2. No extraction required	2. Has relatively low sensitivity and limited range for moisture determinations
	3. More accurate than low frequency resistance of capacitance meters	3. Depends on fluctuation of material density in volume measured
		4. Results affected by factors such as particle size, temperature, soluble salt contents, polarization, and frequency of sample, etc.
E. Dielectric capacitance	1. Has high sensitivity due to large dielectric constant of water	1. Affected by texture of sample, packing, electrolytes, temperature, and moisture distribution
	2. Convenient to industrial operations with continuous measurement system.	2. Potential calibration difficulty beyond pH 2.7-6.7
	3. System can be modified to have universal applicability	3. Difficult to measure “bound” water at high frequencies

Contd...

Table 2.3–Contd...

Method	*Advantgaes*	*Disadvantages*
F. Conductivity	1. Measurement is instantaneous	1. Measures only free water
	2. Non-destructive	2. Conversion charts are needed to obtain total moisture values
	3. Precise	3. Accuracy and precision are affected by temperature, electrolyte content and contact between electrode and samples
		4. Difficult to maintain calibration of equipment
G. Sonic and ultrasonic absorption	1. Bound water can be determined in aqueous solution of electrolytes and non-electrolytes	1. Dependent on type of medium for sound passes
	2. Non-destructive	2. Appropriate standards required to obtain total moisture content
H. Mass spectrometry	1. Can analyze simultaneously a large number of components from complex matrix	1. High variation between theoretical moisture values and hydrated substances
	2. No electrical leakage problem due to low potentials applied to the beam tube	2. Major instrumental problem is memory effect from preceding sample
I. NMR spectroscopy	1. Very rapid analysis	1. Cost of equipment is high
	2. Accurate	2. Separate calibration curves are required for different substances
	3. Non-destructive	3. Constant and correct sample weight required

Contd...

Table 2.3–Contd...

Method	*Advantgaes*	*Disadvantages*
	4. Applicable to many types of foods	4. Not applicable for foods having variable lipid contents
	5. Can differentiate between free and bound water	
	6. Particle size and packing of granular samples have no effect on signal absorption	
J. Neutron-scattering method	1. Density and moisture measured simultaneously	1. Applicable only to substances that are relatively proton-free
	2. Absolute error is claimed to be less than ±0.5 per cent	2. Expensive
	3. Suitable for soil moisture assay.	

yield a close and reproducible estimate of a food's true moisture content.

Several types of vacuum oven have been developed. Laboratory vacuum ovens can be attached to a vacuum line and heated electrically. The front door is typically made airtight by using vacuum grease on a rubber gasket. Although a vacuum of 100-600 mmHg can be maintained inside the sample chamber, it is usually desirable to have pressures below 50 mmHg. The reduced pressure will increase the rate of drying. Dry air is introduced into the vacuum oven during drying; without purging dry air into the oven the vapour pressure of water inside the oven would reduce the usefulness of the vacuum oven, especially for high-moisture foods. The AOAC procedures generally recommend that moisture contents be determined by heating in a vacuum oven at 100°C for 2-6 h at a pressure of 25-100 mmHg.

The general apparatus and procedure for vacuum drying are as follows:

Materials

1. *Vacuum oven*: thermostatically and connected to a vacuum pump maintaining the pressure in the oven, 25 mmHg. The oven should have inlet that passes through an indication and trap for releasing the vacuum.
2. *Dishes*: metal dishes with close-fitting flat bottom to provide maximum contact with the heating plate.
3. Other apparatus and equipment are as those for air-oven drying.

Procedure

- ☆ Wash and dry the metal sample cooling in a desiccator, weigh the dishes.
- ☆ Weigh a 3.0-5.0 g sample into the dish using an analytical balance. Spread the samples evenly over the bottom of dish. Some samples require pre-drying of a dish in an oven to prevent desiccation and splattering.
- ☆ Place the sample dishes in the vacuum oven, partially uncover the dish, evacuate the oven, and dry the sample at an appropriate temperature and vacuum pressure.

During drying, admit a slow current of air dried by passage through indicating desiccant into the oven.

- ☆ Shut off the vacuum pump after 5 h and slowly readmit dry air into the oven. Press cover lightly on to dish using tongs, transfer the dish to a desiccator to cool, and reweigh.
- ☆ Dry for another hour to ensure that constant weight has been achieved.

Calculations

$$\text{Per cent Moisture} = \frac{(\text{loss of weight} \times 100)}{(\text{sample weight})}$$

Per cent Solids = 100–per cent moisture

(*c*) Freeze Drying

The process of freeze-drying or lyophilization is excellent for preserving freshness and textural quality of dried foods. Lyophilization is especially suited for drying high-value liquids, such as coffee and juices, as well as high-value solid foods, like strawberries, shrimp, diced chicken, sliced mushrooms, and even steaks and chops. The process of lyophilization has evolved into a highly advanced drying method. The limitation of this drying process is that its cost may be two to five times greater per weight of water removed than that of other drying methods. Therefore, much of the development work has focused on optimizing both the lyophilization process and equipment to lower drying costs.

The basic principle of lypholization is facilitating the sublimation of water under conditions of reduced pressure and temperature. Sublimation is the direct conversion of ice into water vapor without melting into liquid water; thus, lyophilization preserves the physical structure of the food. Sublimation of ice occurs at temperatures below 0°C and pressure below 4.6 mmHg. Under these conditions, water in the food remains frozen and water vapor leaves the food faster than water in the surrounding atmosphere re-associating on the food, causing a net reduction in the moisture content. Heat is frequently applied to the frozen food to enhance the sublimation rate. The maximum sublimation (*i.e.*, drying) rate occurs when the vacuum is maintained from 0.1-2 mmHg and heat is added just short of melting the ice. As lyophilization proceeds, moisture is initially removed from the surface and progresses towards the center of the frozen food until the final ice sublimes, leaving a moisture content of less than 5 per cent. Drying time may range from 8 h to a few days.

As a moisture analysis method, the application of lypholization methodologies is limited due to the relatively long drying times required and high cost. The primary use of lyophilization would be as a component of a standard reference method for moisture determination.

(*d*) Distillation Methods

Two main types of distillation procedures exist for moisture determination: direct distillation and reflux distillation. In the direct distillation method, a food is heated in a liquid (*e.g.*, mineral oil), which is immiscible with water and has a high boiling point. The water in the food distils directly from this liquid, condenses, and collects in a graduated tube; the volume of the water removed is then measured.

The reflux distillation proceeding is more commonly used than the direct distillation method. Reflux distillation makes use of azeotropic properties of solvent mixtures. During heating, water and an immiscible solvent, such as toluene or xylene, distil off together at a constant ratio and frequently at a temperature lower than the boiling point of either component. For example, the respective boiling points of water and toluene are 100° and 110.6°C, but the boiling point of the binary mixture is 85°C; the distillation ratio of the mixture is approximately 20 per cent water and 80 per cent toluene. If water is denser than its co distillate, as in the case with toluene, the water is again collected in a suitable measuring apparatus where it separates and has its volume measured.

Distillation using a boiling liquid effectively transfers heat to the sample, resulting in rapid distillation. The reduced boiling point of the distillation mixture results in less decomposition of the food during heating. Oxidative reactions are also minimized. This method is especially suited for samples that contain a high concentration of volatile components. The laboratory data obtained from azeotropic distillation have consistently approached the theoretical moisture content to within 0.1 per cent. A comparison between reflux distillation using toluene and oven-drying methods resulted in similar moisture values for a variety of products. The reflux method is not without its potential difficulties (*e.g.*, emulsion formation and suspended water droplets). Using clean glassware and allowing the apparatus to cool before reading the volume of collected water help with these two problems.

The reflux distillation apparatus consists of a heating source under a round-bottom boiling flask. This flask contains the food sample and the solvent (*e.g.*, toluene). A Bidwell-Sterling receiver connects to the round-bottom flask, which will collect and measures the distilled water in a side arm. A condenser is positioned directly above the side arm of the Bidwell-Sterling receiver.

Reagents and Materials

1. Xylene or toluene
2. Reflux distillation apparatus
3. Heating mantle

Procedure

- ☆ Weigh an amount of food so that 2-5 g of water will be released. This could be 10-15 g of cheese or 40 g of spice. Place in an appropriate sized round-bottom flask.
- ☆ Add enough suitable solvent (*e.g.*, toluene cover the food, usually 60-100 ml).
- ☆ Assemble reflux apparatus.
- ☆ Run cold water through the condenser and gradually heat the flask until refluxing start.
- ☆ Adjust the heating to produce two drops condensate per second. When the rate of accumulation decreases, increase heat to four drops per second.
- ☆ When no additional moisture is collected from side arm, rinse the condenser with the solvent and continue heating few more minutes, and heat time is typically 1-1.5 h.
- ☆ Turn off the heat, and allow the apparatus cool, especially the side arm.
- ☆ Record the volume of water in the side arm.

Calculations

$$\text{Per cent Moisture} = \frac{(\text{water volume} \times 100)}{\text{Sample weight}}$$

Detailed experimental protocols for specific products can be found in the Official Methods Analysis of AOAC International.

(e) Chemical Desiccation

This method of moisture determination in dried foods is carried out by desiccation in an evacuated desiccator containing a substance that strongly absorbs moisture. The amount of water removed from the food depends on the strength of the desiccant employed. Table 2.4 lists the relative efficiencies of various desiccating agents. The most effective desiccating agents are phosphorus pentoxide, barium oxide, and magnesium perchlorate. However, phosphorus pentoxide becomes explosive if it absorbs too much moisture. Calcium sulfate is a commonly used desiccant despite it is not being as effective.

Table 2.4: Relative Efficiencies of Selective Desiccants

Substance	*μg Water/L Dry Gas*
P_2O_5	0.02 to 3.6
$Mg(ClO_4)_2$, anhydrous	<0.5 to 0.2
BaO	0.65 to 2.8
H_2SO_4, anhydrrous	3.0
CaO	3.0 to 656
$CaSO_4$, anhydrous	5.0 to 67
Al_2O_3	2.9 to 5.0
$CaCl_2$, anhydrous	67 to 137
$CaCl_2$, granular	140 to 1500
$CuSO_4$, anhydrous	1400 to 2800

Desiccation of the sample usually occurs at room temperature. With few exceptions, desiccation techniques are lengthy procedures, frequently requiring weeks and even months for the sample to achieve constant weight. The equilibrium time depends strongly on the forces holding water in the sample relative to the desiccant. Slight heating may be used in conjunction with the desiccants to enhance moisture removal from the food. Despite its limitations, results obtained using chemical desiccation can serve as reference standards for calibrating more rapid procedures.

(*f*) Thermogravimetric Method

Thermogravimetric analysis (TGA) resembles an automated version of the standard oven-drying method, where moisture in a food is removed via heating. The TGA instrumentation consists of a thermo-balance, which automatically measures and records the weight loss of a food sample as a function of time and temperature while the sample is being heated. A small amount of sample is loaded into the balance, which is then heated under a controlled temperature program. Data are provided in the form of thermogravimetric curves, which plot the sample mass as a function of temperature. Moisture is continuously evaporated from the sample and the weight loss recorded until the sample has reached a constant weight. Because multiple sample transfers are not required, errors associated with sample weighing are minimized. However, as with other methods using heat to determine moisture contents, care must be taken to avoid causing thermal decomposition of the sample which would lead to erroneous results.

Moisture determination using TGA has been shown to give results similar to other methods. The water content of flour as measured by TGA was found to be comparable with that of the Karl Fischer method. The amount of residual moisture in glassy sugars after lyophilization and 1 week storage over P_2O_5 was similar whether determined by TGA, an air oven, or vacuum oven. In addition to total moisture contents, TGA can also be used to quantify chemically bonded water of hydration.

(*g*) Gas Chromatography (GC)

The versatile capability of GC in analytical chemistry can be applied to moisture determination in foods. To analyze a food for moisture using GC, a known mass of the sample is initially homogenized in an anhydrous solvent, such as methanol, ethanol, or isopropanol, resulting in water being extracted into this solvent. The extract is then analyzed using GC such that quantitative separation of the water-solvent mixture can occur. A Poropak Q column and a thermal conductivity detector have been used previously for moisture analysis of meat products. The moisture content is quantified by determining the peak areas of water and solvent; these areas then compared to the areas of solutions containing known amounts of water (*i.e.*, a standard curve). A comparison of moisture values obtained for meat samples using GC and either distillation or oven-drying methods indicated that the GC values were not different from those obtained using the more conventional methods. Analysis is rapid, but requires specialized requirement.

(*ii*) Indirect methods

As mentioned previously, a variety of indirect methods for moisture determination also exist. These methods measure a property of water that is dependent on its content. Thus, moisture contents can be calculated using appropriate calibration curves.

(*a*) Refractometry

This optical method measures the refractive index of a solution, which can then be used for determining its moisture content. The refractive index increases as the moisture content deceases. By measuring the refractive index of a solution or slurry, the moisture content can be rapidly determined using an appropriate calibration curve. Refractory is best suited for high-sugar products, such as fruit products and syrups. An official method based on refractive index measurement exists for the analysis of moisture in honey.

For solid or semisolid foods, the sample can be homogenized with an anhydrous solvent (*e.g.*, isopropanol) and then the solution's refractive index can be measured using a refractometer. A calibration curve is produced by measuring the refractive index of solutions containing the same solvent with known amounts of added water. The moisture content of the sample is required and strict temperature control is necessary because the refractive-index measurement is temperature sensitive. Although moisture can be determined using refractometry, its more common use is for the determination of soluble solids.

(*b*) Infrared Absorption Spectrometry

Moisture content can also be determined using infrared (IR) spectroscopy. The IR spectrum of a chemical compound has been described as one of its most characteristic physical properties. Because of this, IR spectroscopy is one of the most versatile methods for determining the moisture content of a large variety of substances in solids, liquids or gases by employing suitable wavelengths at which maximum absorption is expected to occur. For water, the spectral region of interest is 700-2400 nm; absorption bands occurring at 1450 and 1940 nm are frequently used. The moisture content in a sample can be determined by comparing the band intensity with that of the same band for standard concentrations of water. IR methodologies have been used to determine the moisture content of grains and fruits and vegetables. An official method employing IR exists for moisture determination in dried vegetables.

The basic concept of this methodology is that an IR beam passes through an optical filter, which consequently transmits energy at a specific wavelengths through the sample cell and then to a detector. Ideally the wavelength used is that of maximum absorption for the compound being measured. The IR spectrum must be calibrated using standards of known concentration. Comparisons can then be made between the absorption from the sample with those of the standards.

Another approach has been employed for the analysis of fluid milk. The total solids of milk can be determined using IR. In milk, wavelengths of 5723, 6465, and 9610 nm are used to determine the fat, protein, and lactose contents, respectively. By using an appropriate calibration technique total solids can be determined. The moisture content would be calculated by subtracting the total solids value from 100.

(*c*) Near-Infrared (NIR) Reflectance Spectroscopy

The high resolving power of reflectance spectra in the NIR range (800-2500 nm) was recently utilized to develop NIR reflectance spectroscopy. The mid-IR range (2.5-24 μm) has high resolution in the absorption spectrum and can absorb IR radiation effectively from many compounds, but resolution of the reflectance spectrum is poor.

The speed of analysis is the primary advantage of the method. The accuracy of the NIR reflectance method depends on the calibration curve, derived from wet chemical analysis of the standard samples. If the calibration sample set does not adequately represent the range of the unknown samples, then the analysis will be error prone. Linear calibration curves (correlation coefficients > 0.98) were shown for moisture in raw pork and beef as determined by drying and NIR reflectance spectroscopy.

The reflectance (R) spectra from an IR spectrophotometer with a monochromator operated in a single or double-beam mode can be downloaded into a computer. Ground samples are packed into a sample holder maintaining direct contact with a concentric IR-transmitting quartz window. The reflected radiation signals of the diffused spectra from the glass window are collected with four lead sulfide detectors equally spaced around the incident beam. The signals from the detectors are amplified with a logarithmic response amplifier, digitized, and fed into a computer. The wave length range from approximately 1100 to 2500 nm is scanned every 2 nm (or 0.5 nm) width of the reflectance curve. Both the IR reflectance (R) curve and the log (I/R) curve can be recorded as the second derivative of the original curves to help evaluate overlapping absorption bands.

Reflectance data cannot be used directly for quantitative analysis. Moisture contents of standard samples are imputed into the computer along with the NIR data, and these are analyzed with a stepwise multiple linear-regression method to develop prediction equations by a regression analysis of NIR spectral data against chemical data. For the moisture determination of the raw meat samples, this spectral data consisted of taking the ratio of the second derivative of the log (I/R) data at two different wavelengths demonstrating the complexity of the analysis.

(*d*) Microwave Absorption Method

Water absorbs several thousand times more microwave energy than a similar amount of a dry substance. For example, at frequencies between 1-30 GHz, the loss tangent of water is 0.15-1.2 whereas for dry materials the loss tangent is 0.001–0.05. Due to these differences, the absorption of microwaves can be used to determine the moisture content of a variety of food products. The absorption of microwave energy at 2450 MHz increases linearly with increasing moisture content.

The microwave moisture meter consists of a constant source of microwave radiation, a wave guide, a detector, a microwave attenuator and amplifier, and an indicating meter. The sample is placed between a microwave transmitter and receiver. The difference in attenuation readings between the transmitter and receiver is the attenuation of the sample, which is dependent on moisture and is therefore used to construct a calibration curve. The accuracy of the microwave measurement is affected by various factors, including leakage of microwave energy, sample temperature, particle size, polarization of different material and the presence of soluble salts. The device was used to determine the moisture content of cakes and was found to give results similar to those from using a drying oven.

(*e*) Dielectric Capacitance

The dielectric constant of water at 20°C is around 80, whereas those of fatty acids and sucrose are both around 3. A 1 per cent increase in moisture content of a substance will theoretically increase its dielectric constant by approximately 0.8. Furthermore, the dielectric constant of water-containing substances increases almost linearly up to approximately 30 per cent moisture content. Based on this information, instruments utilizing dielectric measurements for moisture determination were developed.

The central component of the dielectric instrument is a capacitance cell, consisting of two metal plates spaced apart, which have equal, but opposite charges. These charges reverse at fixed frequencies to yield an alternating current. A sample placed between the two plates will become polarized and change the capacitance of the plates. This capacitance change, which is affected by moisture content, is measured by the instrument. Upon calibration with standards of known moisture content, moisture contents of food samples can be determined. Moisture analysis using a capacitance meter is influenced by moisture distribution, presence of electrolytes, temperature, and sample density.

(*f*) Conductivity Method

In conductivity methods, conductivity and resistance are measured in an electrical circuit containing a food sample. There is a distinct relationship between the moisture content of materials and their electrical properties. Electrical resistance decreases and conductivity increases as the moisture content of a food increases. Measuring resistance appears to be most useful. Frequently, the logarithm of resistance is plotted as a function of either moisture content or humidity. The logarithm of resistance is basically linear as a function of humidity, but shows curvature as a function of moisture content.

Various instruments exist for measuring either resistance or conductance of foods. Proper calibration can result in the accuracy of this method being less than ±0.5 per cent. For determining moisture using a conductivity meter, the food is placed between two electrodes and the current flowing through the sample is measured by the change in electrical resistance. Conductance readings are converted into moisture contents using a table that corrects for sample temperature. Conductivity methods measure only the "free" water in the sample; the amount of "bound" water needs to be added to approximate more closely the total moisture content. Conversion charts may be required due to the variation of "bound" water among samples. Like the previously discussed dielectric capacitance determination, the accuracy and precision are affected by moisture distribution, temperature, and electrolyte content as well as the quality of contact between the electrodes and samples.

(*g*) Sonic and Ultrasonic Absorption

The absorption of sound energy depends on the type of medium through which it is transmitted. Thus, as the moisture content of the medium changes, so too does the amount of sonic and ultrasonic absorption. Based on this principle, ultrasonic velocity measurements have been used for the determination of moisture content. In addition to absorption, the high frequency of ultrasonic waves also permits their reflection and refraction.

The food sample to be analyzed is positioned between an energy generator and microphone. The energy out put of the sample is amplified, yielding a voltmeter reading. By using an appropriate standard curve, voltmeter readings are converted into moisture contents. Ultrasonic methods have been used in laboratory settings to determine the composition including moisture content of chicken and dry fermented sausages. This technique was found to be rapid and non-destructive, while giving results similar to those of standard methods.

(*h*) Cryoscopic Methods

Another indirect method for determining moisture utilizes the colligative property known as freezing point depression. The freezing point of water decreases as the concentration of dissolved solutes increases. Thus, for liquids containing a constant type of solute, the freezing point inversely correlated with the amount of water present. The most common use of this method is to measure for water added to fluid milk. The freezing point of "natural" milk is approximately–0.54°C; a freezing point closer to 0°C indicates water has been added to the milk (*i.e.*, has become adulterated). Cryoscopic measurements have also been successfully used to evaluate the dissolved solids content of various fluid dairy products. Moisture could then be estimated by subtracting the dissolved solids content and fat content from 100.

(*i*) Other Indirect Instrumental Methods

Several other methods exist that require substantial instrumentation that may be applied to the determination of moisture in foods. These methods include nuclear magnetic resonance, mass spectrometry and neutron scattering.

Nuclear magnetic resonance (NMR) is a quick and non-destructive method of moisture determination. Instead of utilizing properties of water, this technique utilizes the nuclear properties of its protons. The challenge is to differentiate the proton NMR signal of water from the other hydrogen containing substances in the food. NMR techniques have been used to determine the moisture content in oilseeds as well as various grains. The accuracy of NMR techniques is approximately 0.2 per cent.

Moisture analysis using mass spectrometry has also been described. Although mass spectrometry has been widely utilized to identify unknown substances in food samples, the quantification of water is more problematic. The most significant problem is the "memory effect" where the previously analyzed sample influences the results of the following sample. Some success has been obtained by reacting cryogenically concentrated moisture with calcium carbide to yield acetylene, which is then quantified. However, mass spectrometry remains an uncommon method for moisture determination.

When energized neutrons interact with nuclei, the neutrons scatter. Hydrogen atoms are the most effective at scattering neutrons, which therefore forms the basis of a determination of moisture using neutron-scattering methods. As with the NMR method discussed previously, neutrons will scatter from any hydrogen nucleus, not just those associated with water. Thus, neutron scattering methods are most appropriate for samples low in non-aqueous protons, such as inorganic substances. Because food is primarily composed of protein-rich organic material, the application of neutron scattering to moisture determination of food is virtually non-existent.

(*iii*) Methods of Moisture Estimation

Preparation of Sample (AOAC, 1975)

Separate meat completely free from bone, pass rapidly three times through chopper with plate opening of 3 m diameter, mix thoroughly after each grinding and begin all determinations promptly. If any delay occurs, chill sample to inhibit decomposition. Keep ground material in glass or similar containers with air and water-tight covers.

Method–I

Chemical Methods and their Application

The application of chemical analytical methods to the maintenance of compositional uniformity has been greatest contribution of the scientists to quality control

Moisture

The following method is simple and rapid and will results that are within 3 per cent of AOAC methods. The method is adapted from the American Meat Institute Foundation (Everson *et al.*, 1955).

Reagents and Materials

1. Capryl alcohol (2-octanol)
2. Distillation assembly
3. Weighing balance

Procedure

- ✰ Assemble distilling apparatus consisting of a round bottom 250 ml distillation flask with standard taper joints and heated by an electric heating mantle controlled by a variable transformer, a Leibig condenser and a Dean and Stark 10 ml receiver graduated to 0.2 ml divisions.
- ✰ Weigh 15 g ± 0.1 g of the well ground and mixed sample on a small piece of tarred cheese cloth and immediately transfer to the distillation flask
- ✰ Add100 ml of Capryl alcohol (2-octanol)

- ☆ Heat the distillation flask and continue distillation at a fairly rapid rate until the water layer (bottom) in the receiver remains constant for 4 min.
- ☆ Read the volume of water in the receiver and calculate percentage of moisture as

$$\text{Per cent Moisture} = \frac{\text{Volume of water} \times 100}{\text{Weight of sample}}$$

- ☆ For products containing over 65 per cent moisture, use only 10 g sample and substitute 10 for 15 as denominator in the above-mentioned equation.

Note: Capryl alcohol has an irritating odour and should be used in a good or well-ventilated location

The above moisture analysis takes only 15 min and can often be completed more rapidly. More accurate results can be obtained by using tetrachlorethylene as the solvent and distilling 15g sample for 1h. This system should give results, which are 1 per cent of AOAC methods and is recommended where time permits. It is, in fact, preferable to the official method on several counts.

Method–II

Determination of Moisture

Materials

1. Aluminium dish (Moisture cups) with lid
2. Dessicator
3. Hot air oven
4. Weighing balance

Procedure

- ☆ With lids removed, dry the sample containing about 2 g dry material 16-18 hours at 100-102°C in hot air oven.
- ☆ Use covered aluminium dish >15 m diameter and > 40 m deep.
- ☆ Cool in desiccator and weigh.
- ☆ Report loss in weight as moisture (AOAC, 1975).

Method–III

Determination of Moisture

Reagents and Materials

1. Ethanol
2. Glass rod
3. Weighing balance
4. Aluminium dish (Moisture cups) with lid
5. Dessicator
6. Hot air oven

Procedure

- ✰ Dry the dish containing a quantity of sand 3 or 4 times the mass of the test portion and the glass rod for 30 minutes in the oven at 103±2°C.
- ✰ Allow the dish with its content to cool in the desiccator to room temperature and weigh to the nearest 0.001 g.
- ✰ Transfer 5 to 10 g of the prepared sample to the dish and weigh the dish again to the nearest 0.001 g.
- ✰ Add 5 to 10 ml of ethanol depending on the mass of the taste portion and mix the mass by means of the glass rod.
- ✰ Place the dish on the water bath regulated at 60-80°C and heat until ethanol has evaporated, stirring occasionally. Heat the dish and contents for 2 hours in the hot water bath at 102±2°C.
- ✰ Cool to room temperature and again weigh (IS:5960, Part V-1971).

$$\text{Moisture content (per cent)} = \frac{W_1 - W_2}{W_1 - W_0} \times 100$$

where,

W_1 = Weight of dish with test material before drying

W_2 = Weight of dish with test portion after drying

W_0 = Weight of dish, rod and sand.

Method–IV

Determination of moisture by Air Oven method

Principle

The sample is dried to constant weight in an air oven. The method is applicable to all food products except those that may contain volatile compounds other than water or those liable to decompose at 100°C.

Materials

1. *Oven*: Temperature 100-102°C.
2. *Dishes*: Nickel, stainless steel, aluminium or porcelain. Metal dishes should not be used when the substance to be dried may produce corrosive action.
3. *Desiccator*: Containing dry phosphorus pentaoxide, calcium chloride or granular silica gel.

Procedure

- ☆ Dry the empty dish and lid in the oven for 15 minutes at 100°C and transfer to the desiccator to cool for about 10-20 minutes.
- ☆ Mix the prepared sample thoroughly and transfer about 5 g to the dish. Replace the lid and weigh the dish and contents as rapidly as possible to the nearest mg.
- ☆ Remove the lid and place the dish and lid in the oven avoiding contact of the dish with the wall. Dry for 6 hours at 100°C. For products that do not decompose during long period of drying, it is permissible to dry overnight that is about 16 hours.
- ☆ Remove the dish from the oven, replace the lid, cool in desiccator and reweigh when cold.
- ☆ Dry for a further hour to ensure that constant weight has been achieved.

Calculations

Let weight (g) of sample= W_1

Loss of weight (g) = W_2

Weight of dried sample = W_3

Then, $$\text{Moisture (per cent)} = \frac{W_2 \times 100}{W_1} \times 100$$

$$\text{Total solids (per cent)} = \frac{W_3}{W_1} \times 100$$

Method–V

Determination of Moisture by Vacuum Oven Method

Principle

Many products that decompose in an oven at 100°C can be dried at a lower temperature under reduced pressure. The efficiency of the method depends on maintaining as low a pressure as possible in the oven and on removing water vapours from the oven quickly.The method is applicable to food products that contain compounds liable to decompose at 100°C.

Materials

1. *Vacuum Oven*: Thermostatically controlled and connected through a drying train to a vacuum pump capable of maintaining the pressure in the oven below 25 mm of Mercury.
2. *Dishes*: Metal dishes with close fitting lids and flat bottom to provide maximum area of contact with the heating plate.
3. *Desiccator*: Containing dry phosphorus pentaoxide, calcium chloride or granular silica gel.

Procedure

- ✰ Dry, cool and weigh a dish and lid to the nearest mg.
- ✰ Transfer to dish and spread out about 5 g well mixed sample. Replace the lid and weigh the dish and contents as rapidly as possible to the nearest mg.
- ✰ Partially uncover the dish and place in the oven, evacuate the oven and dry the sample at 70°C for 6 hours. If the material contains an appreciable amount of water, it is an advantage to pass a slow flow of dry air into the oven for the first 1-2 hours.

- ☆ After the period of heating, admit a slow season of dry air to the oven until atmospheric pressure is reached, cover the dish, transfer to the desiccator to cool and reweigh when cold.
- ☆ Dry for a further hour to ensure that constant weight has been achieved.

Calculations

Let weight (g) of sample = W_1

Loss of weight (g) = W_2

Weight (g) of dried sample = W_3

Then,

$$\text{Moisture (per cent)} = \frac{W_2}{W_1} \times 100$$

$$\text{Total solids (per cent)} = \frac{W_3}{W_1} \times 100$$

Method–VI

Determination of Moisture by Dean and Stark Distillation Method

Principle

A known quantity of sample is heated with toluene or xylene, because:

(*a*) It is immiscible with water.

(*b*) There is significant difference in density of toluene and water.

(*c*) Volatile acids are soluble in toluene. The water presents in the sample evaporates. It is condensed by the condenser. Since heavier than toluene, water collects at bottom of graduated tube by which moisture per cent in sample may be determined.

The method is applicable to fatty foods and those containing significant volatile substances other than water.

Reagents and Materials

1. Xylene or Toluene.

2. Dean and Stark distillation apparatus.
3. Heating mantle.

Procedure

- ☆ Weigh 10-20 g of the sample and place in a 250 ml flask.
- ☆ Add 100 ml of xylene or toluene to the flask.
- ☆ Connect the flask to the condenser, heat the flask and contents on the heating mantle and bring the contents to the boil.
- ☆ Adjust the heating mantle so that the flask contents are just kept boiling and continue heating for at least 1½ hours.
- ☆ Switch off the heating mantle and allow the apparatus to cool.
- ☆ Record the volume of the water in the side arm.

Observations

Volume of water collected = x ml

Weight of the sample = y gm.

Calculations

$$\text{Moisture per cent} = \frac{x}{y} \times 100$$

Notes

- ☆ It is good for volatile substances like spices, ginger, cardamom for moisture determination.
- ☆ It avoids loss of volatile, hence gives better result.
- ☆ Disadvantages of toluene distillation method is water get accumulated around the condenser.

Method–VII

Determination of Moisture by Karl Fischer Method

Principle

The water in the sample is titrated by means of Karl Fischer reagent which consists of SO_2, pyridine and iodine in anhydrous methanol. The reagent is standardized against the water of crystallization of hydrated sodium acetate. The end of the titration is detected electrometrically, utilizing the dead stop end point technique.

This method is used to determine moisture content in dehydrated products.

Reagents and Materials

1. Methanol–Anhydrous for Karl Fisher titration contains 11 per cent pyridine.
2. Sodium acetate trihydrate
3. Karl Fisher reagent : The reagent is standardized daily by the use of sodium acetate in place of the sample.
4. Burette: All glass, automatic filling type fully protected against moisture.
5. Electrometric apparatus.
6. Titration vessel.

Standardization of Karl Fischer Reagent

The water content (M per cent) of hydrated sodium acetate is accurately determined by oven drying at 120°C for 4 hours.

1. Weigh to the nearest mg approximately 0.4g of hydrated sodium acetate into a pre-dried 50 ml round bottom flask.
2. Pipette 40 ml methanol into the flask and stopper immediately.
3. Swirl the contents until the added sample is completely dissolved.
4. Titrate 10 ml aliquot of this solution and 100 ml of methanol blank as described in the procedure.

Procedure

- ☆ Weigh to the nearest mg, an amount of sample containing approximately 100 mg water into a pre-dried round bottomed 50 ml flask.
- ☆ Pipette 40 ml of methanol into the flask quickly. Place it on the heating range and connect the reflux condenser.
- ☆ Gently boil the contents of the flask under reflux for 16 minutes.

- ☆ Remove the flask from the heat and with the condenser attached, allow to drain for 15 minutes.
- ☆ Remove and stopper the flask.
- ☆ Pipette a 10 ml aliquot of the extract into the titration vessel and titrate with Karl Fischer reagent to the dead stop end point and record volume of titrant used.
- ☆ Determine a blank titre by taking 10 ml aliquot from 40 ml methanol, which has been refluxed as has been described above.

Calculations

(*a*) Water equivalent of Karl Fischer reagent:

Water content of sodium acetate (per cent) = M

Weight (g) of sodium acetate trihydrate = W

Volume (ml) of titrant used for standard = Vs

Volume (ml) of titrant used for blank = Vb

Then the water equivalent of Karl Fischer reagent F (mg water per ml)

$$= \frac{W \times M \times 2.5}{Vs - Vb}$$

(*b*) Determination of water content in sample:

Let weight of sample (g) = W_1

Volume of reagent used for sample (ml) = V_1

Volume of reagent used for blank (ml) = V_2

Standardization factor of reagent (mg water per ml) = F

Water content (per cent) of sample

$$= \frac{0.4 \times F \times (V_1 - V_2)}{W_1}$$

Precautions

- ☆ Use redistilled methanol using a small quantity of magnesium and a few crystal of iodine.
- ☆ All titrations should be made quickly with as little delay on possible between addition of reagents.

Method–VIII

Determination of Moisture Content (Sp : 18 Part XII)–1984. ISI Hand Book of Food Analysis.

Principle

Thorough mixing of the test portion with sand and ethanol, pre-drying of the mixture on a water bath and drying to constant mass at 103±2°C.

Reagents and Materials

1. Sand: Use the fraction of then sand which passes through a 1.5 mm IS sieve and stays on a 0.25 mm IS sieve. Wash the sand with running water. Boil the sand with dilute hydrochloric acid (dil HCl) for 30 minutes while stirring continuously. Repeat this with another portion of the acid until the acid no longer turns yellow after boiling. Then wash the sand with distilled water until the test for chloride is negative. Dry the sand at 150 to 160°C and store in an air tight closed bottle.
2. Ethanol–95 per cent (V/V).
3. Meat mincer–Lab. size fitted with a plate with holes of diameter not exceeding 4 mm.
4. Dish, Flat–of porcelain or metal. (*e.g.* nickel, aluminium, stainless steel), diameter at least 60 mm., height about 25 mm.
5. Thin glass rod : flattened at one end, slightly longer than the diameters of the dish.
6. Drying oven–electrically heated, adjusted to operate at 103±2°C.
7. Water bath
8. Desiccator–containing an efficient desiccant.
9. Analytical balance.

Procedure

Preparation of Sample: Proceed with a representative sample of at least 200g. Render the sample uniform by passing it at least twice through the meat mincer and mixing. Keep it in a completely fitted, air-tight container and store in such a way that the deterioration and change in composition are prevented. Analyze the sample, as soon as possible but in any case within 24 hours.

Test Portion: Dry the dish containing a quantity of sand three or four times the mass of the test portion and the glass rod for 30 minutes in the oven at 103±2°C. Allow the dish with its contents to cool in the desiccators to room temperature and weigh to the nearest 0.001g. Transfer 5 to 10 g of the prepared sample to the dish and weigh the dish again to the nearest 0.001 g.

Determination: Add 5 to 10 ml of ethanol depending on the mass of the test portion and mix the mass by means of glass rod. Place the dish on the water bath regulated at a temperature between 60°C and 80°C in order to avoid the ejection of particles, and it until the ethanol has evaporated, stirring occasionally. Heat the dish and contents for 2 hours in the drying oven regulated at 103±2°C. Remove the dish from the oven and place it in the desiccator.

Allow the dish to cool to room temperature and weigh it to the nearest 0.001 g. Repeat the operations described above until the results of two successive weighings, separated by 1 hour's heating do not differ by more than 0.1 per cent of the sample weight. Carry out two determinations on the same prepared sample.

Observation

Mass, in g, of the dish containing the test portion, rod and sand, before drying = M_1

Mass, in g, of the dish containing the test portion, rod and sand, after drying = M_2.

Mass, in g, of the dish, rod and sand = M_0.

Calculations

Moisture content, as a per cent by weight is equal to:

$$= \frac{M_1 - M_2}{M_{1-}M_0} \times 100$$

Calculate the arithmetic mean of the two determinations. Report the result rounded to one decimal place.

Repeatability

The difference between the results of two determinations carried out simultaneously or in a rapid succession by the same analyst should not be greater than 0.5 g of moisture per 100 g of sample.

Method–IX

Determination of Moisture Content (Kramlich *et al.*, 1973)

Principle

Determination of moisture is made by drying the sample at elevated temperatures. Per cent moisture is derived from the difference in weight of the sample before and after drying.

Materials

1. Analytical balance
2. Drying oven.
3. Aluminum cups with covers, 2½ in. diameter, 3 $^3/_4$ in. deep.
4. Desiccator.
5. Laboratory grinder, chopper, or blender.

Procedure

- ✰ Select a representative product sample. If not already finely communicated, grind, chop, or communicate in a blender until finally divided and uniform in composition.
- ✰ Weigh exactly 30.0 gm of the chopped sample into a previously weighed aluminum can.
- ✰ Dry in an oven at 212°F to a constant weight (about 12 to 16 hour).
- ✰ Cover the can and allow to cool in a desiccator before weighing.

Calculations

$$\text{Weight of solids} = \begin{array}{c}\text{Weight of dried sample}\\ \text{and container}\end{array} - \begin{array}{c}\text{Weight of}\\ \text{container}\end{array}$$

$$\text{Per cent solids} = \frac{\text{Weight of solids}}{\text{Weight of original sample (300 gm)}} \times 100$$

Per cent moisture = 100.00–per cent solids.

Test Limitation

☆ The method is dependent on the volatilization and subsequent evaporation of water from the sample. The meat sample may contain substances other than water that are volatile.

2. Protein Estimation

Protein estimation is broadly classified into i). Total nitrogen determination followed by conversion to crude protein content and ii) direct protein determination. The above classification along with the corresponding determination methods are mentioned below at Table 2.5 and the advantages and disadvantages of different methods are given at Table 2.6.

Table 2.5: Methods for Total Protein Determination in Food and Food Products

Method Classification	*Determination Method*
Total nitrogen determination, followed by conversion to crude protein content	Kjeldahl
	Dumas/pregl-Dumas
	Neutron activation
	Proton activation
Direct protein determination	Formal titration
	Direct distillation, Kofranyi
	UV spectrophotometry
	UV absorption
	UV fluorescence
	Visible-region spectrophotometry (colorimetry)
	Biuret
	Folin-Ciocalteau/Lowry
	Bicinchnoic acid (BCA)
	Dye binding
	Silver staining

Contd...

Table 2.5–Contd...

Method Classification	Determination Method
	IR Spectrophotometry
	IR absorption
	NIR transmittance
	Pulsed nuclear magnetic resonance (NMR) spectrophotometry
	X-ray photoelectron spectrophotometry
	Turbidimetry/nephelometry
	Refractometry
	Polarography
	Radioactivity(liquid scintillation counter)
	Photoacoustic spectroscopy

Table 2.6: Advantages and Disadvantages of Methods Used for the Determination of Total Protein in Food and Food Products

Method	Advantages	Disadvantages
Kjeldahl	Approved for various (soluble and insoluble) types of food products High reliability and accuracy Included in methods approved by international organizations	Interference by nonprotein nitrogen compounds Excessive foaming during digestion Use of toxic and/or expensive catalysts and generation of waste products Choice of conversion factor Low sensitivity Time consuming (2h), length of after-boiling time
Dumas	No Hazardous waste generated Improved accuracy and reliability Short analysis time (3 min)	Expensive equipment Measures total mitrogen
Biuret	No interference by free amino acids Little influence of amino acids composition on color development Simplicity of operation, ease of handling large number of samples	Internecine from ammonia, detergents, and buffer, salts, *e.g.* Tris Low sensitivity (concentration: 1-6 mg: ml^{-1}; amount of protein needed 0.05-5 mg)

Contd...

Table 2.6–Contd...

Method	*Advantages*	*Disadvantages*
Lowry	High sensitivity (amount of protein needed +5-10 ug) Simplicity of operation, ease of handling large number of samples.	

Method–I

Estimation of Protein Content (Kramlich *et al.,* 1973)

Principle

Determination of protein in a meat sample is done by measuring total nitrogen in the sample by the standard Kjeldahl method and converting this value to per cent protein.

Reagents and Materials

1. 35 cc. Concentrated sulfuric acid.
2. 15 g potassium sulfate AR
3. 400 ml distilled water
4. 60 ml 50 per cent sodium hydroxide
5. 50 ml 2 per cent boric acid
6. Alcoholic solution(5 parts 0.1 per cent Bromocresol Green + 1 part 0.1 per cent Methyl Red)
7. 4 g Mossy zinc (C.P.)
8. 0.5 N Hydrochloric acid.
9. Kjeldahl digestion assembly.
10. Kjeldahl distillion apparatus
11. 2,800-cc. Kjeldahl flasks
12. 1,100-cc. Graduate cylinder
13. 2,500-cc. Erlenmeyer flasks.
14. 1,50 cc burette
15. Analytical balance
16. 2 pieces of toned nitrogen free filter paper.
17. 4 glass beads.
18. Copper wire (# 18 gauge, 3 inches length)

Procedure

- ✰ Prepare the sample by grinding the fat free residue (see analysis of fat).
- ✰ Weigh an aliquot of sample on a piece of tared nitrogen free filter paper. Carefully fold the filter paper containing the sample and transfer to a Kjeldahl flask.
- ✰ Add a piece of copper wire, 35 ml sulfuric acid, 2 glass beads and 15 g potassium sulfate.
- ✰ Heat the mixture gently on the digestion apparatus until frothing ceases.
- ✰ Boil briskly, and continue the digestion for a time after the mixture is colorless (about 2 hours).
- ✰ Cool the flask. Slowly add 400 ml distilled water and 60 ml 50 per cent sodium hydroxide. Pour the sodium hydroxide solution down the side of the flask so that it does not mix at once with the digest.
- ✰ Add a chunk of the massy zinc to the flask.
- ✰ Transfer the Kjeldahl flask to the distilling apparatus and connect it to the condenser by means of the Kjeldahl connector.
- ✰ Place the condenser up in a 500 ml Erlenmeyer receiving flask containing 50 ml 2.0 per cent boric acid solution.
- ✰ Mix the contents by shaking heat gently, and distil 150 to 200 ml of distillate into the receiving flask.
- ✰ Break contact of the condenser tap with the distillate and continue distillation 2 to 5 min. to steam out the condenser.
- ✰ Titrate the distillate with standardized 0.5N hydrochloric acid.
- ✰ The end point is reached when the distillate color changes from blue green to colorless.

Calculations

$$\%\ \text{protein} = \frac{\text{ml Hydrochloric acid} \times \text{normality} \times 0.014 \times 6.25 \times 100}{\text{Weight of aliquot sample}}$$

If numerous Kjeldahl determinations are to be made, a practical modification is an adjustment of the normality of the standardized hydrochloric acid to 0.5>143 N. The protein factor (normality × 0.014 × 6.25) is then 0.0500. If the weight of the sample aliquot is calculated at 1/10 of the percent of fat free residue, percent protein is calculated as follows:

$$\frac{\text{ml of titer}}{2}$$

Test Limitations

It is assumed that all the nitrogen is found in the proteins and that the nitrogen content of the protein is 16 per cent. Therefore, a factor of 6.25 is used.

Precautions

- ☆ Kjeldahl digestion should be carried on under a fumeless hood.
- ☆ Exercise care when handling strong acids.
- ☆ Wear safety glasses.

Method–II

Determination of Total Protein by Micro-Kjeldahl Distillation method (AOAC, 1975)

Principle

All the nitrogen compound of meat converted to ammonium sulphate by boiling with concentrated H_2SO_4. Subsequently the ammonium sulphate was hydrolysed by strong alkali. By steam distillation the liberated ammonia was collected in boric acid solution containing "Toshiro's indicator". Ammonia with boric acid form Ammonium borate which was then titrated against standard H_2SO_4.

Reagents and Materials

1. Sulphuric acid (Concentrated) AR
2. NaOH–40 per cent
3. Digestion mixture
4. Standard H_2SO_4–N/70
5. Toshiro's indicator

6. Kjeldahl flask
7. 250 ml volumetric flask
8. 50 ml burette
9. 100 ml conical flask
10. 5 or 10 ml pipette
11. Digestion bench
12. Micro-Kjeldahl distillation apparatus
13. Glass beads
14. Measuring cylinder 20 ml.

Digestion mixture (9.5 part Sodium Sulphate + 0.5 parts Copper Sulphate)

Toshiro's indicator

Methyl red = 80 mg

Bromocresol green = 20 mg

Methyl alcohol = 100 ml

10 ml of above solution added to 1.0 litre of 2 per cent boric acid solution.

Procedure

- ☆ Weigh 2.0 g of meat in a butter/filter paper and transfer it to a Kjeldahl flask.
- ☆ Add 20 ml of concentrated H_2SO_4 and a pinch of digestion mixture (which will act as a catalyst).
- ☆ Digest the sample on digestion heater until the solution turns clear green.
- ☆ Transfer the digested sample into 250 ml volumetric flask with distilled water.
- ☆ Take 10 ml of aliquot into Micro-Kjeldahl distillation assembly and add 20 ml of 40 per cent NaOH.
- ☆ Take 10 ml of Toshiro's indicator in a conical flask and dip the tip of silver tube into it.
- ☆ Heat the alkaline liquid by passing steam into it until boiling and keep it so till 30 ml of distillate is collected in the conical flask.

- ☆ Remove the conical flask after rinsing the tip with a little distilled water.
- ☆ Titrate the distillate with standard N/70 H_2SO_4 to a light pink end point. Always perform a blank test and correct the readings accordingly.

Calculations

$$\% \text{ total Protein} = \frac{\text{Amount of acid consumed } (N/70\ H_2SO_4) \times 0.0002 \times 250 \times 6.25}{\text{Weight of sample} \times \text{ml of aliquot taken}} \times 100$$

Method–III

Estimation of Crude Protein by Kjeldahl Method

Principle

When an organic matter is digested with concentrated H_2SO_4, organic nitrogen is converted into inorganic nitrogen or salt of nitrogen, say $(NH_4)_2SO_4$. The nitrogen in ammonical salt can be determined by distillation with an alkali in a Kjeldahl's apparatus. The solution of the salt is distilled with 50 per cent sodium hydroxide (NaOH) and ammonia (NH_3) thus liberated is absorbed in a known excess of 4 per cent boric acid solution. The amount of ammonia absorbed is determined by titrating it with standard acid (N/10 HCl or H_2SO_4) using methyl red and bromocresol green as indicator. The end point is indicated by the change of color from blue to pink.

Reactions

$$\text{Organic N} + \text{concentration } H_2SO_4 \longrightarrow (NH_4)_2SO_4 + H_2O + CO_2\uparrow$$

$$(NH_4)_2SO_4 + 2NaOH \longrightarrow Na_2SO_4 + 2NH_3\uparrow + 2H_2O$$

$$2\,NH_3 + 2HCl \longrightarrow 2\,NH_4Cl.$$

Reagents and Materials

1. N/10 Hydrochloric acid (HCl)
2. 4 per cent Boric acid
3. Mixed indicator
4. 40 per cent Sodium hydroxide (NaOH)
5. Digestion mixture

6. Kjeldahl's digestion assembly
7. Kjeldahl's distillation apparatus
8. Burette
9. Pipette 2 ml, 10 ml
10. Titration flask 100 ml

4 per cent Boric acid: Dissolve approximately 40 g of boric acid in 1 litre of distilled water. Add 5 ml of mixed indicator and adjust the stock solution by adding diluted HCl so that it takes slight pink color.

Mixed indicator: Take 0.5 g of bromocresol green and 0.1 g of methyl red and dissolve in 100 ml of 95 per cent ethanol, adjust this solution with diluted NaOH or HCl to bluish purple color.

40 per cent Sodium hydroxide (NaOH): Dissolove approximately 400g of caustic soda (NaOH) in 1 litre of water.

Digestion mixture: 10g potassium sulfate (K_2SO_4), 0.5g copper sulfate ($CuSO_4$) and 1 g of ferrous sulfate ($FeSO_4$) are mixed well. Here potassium sulfate acts by raising boiling point of H_2SO_4 and copper sulfate acts as catalyst to promote oxidation of protein in the sample.

Procedure

1. Digestion

- ☆ Exactly weigh 0.5g of the given sample and transfer to Kjeldahl's digestion flask (100 ml).
- ☆ Add 10 ml of concentration H_2SO_4 to it and also mix pinch of digestion mixture.
- ☆ At first gently heat, then more strongly after frothing has ceased.
- ☆ Stop heating, when the solution becomes clear or colorless.
- ☆ Cool the solution.
- ☆ Dilute the solution with distilled water to make the volume 100 ml.

2. Distillation

- ☆ Before starting the actual distillation with the given sample, take blank reading to avoid any nitrogen retained in the apparatus.

- ☆ Pipette out 10 ml of 4 per cent boric acid solution into a 100 ml conical flask, add 2 drops of mixed indicator to it and fit the conical flask to the distillation apparatus.
- ☆ Take 2 ml of digestion solution and put in the distillation apparatus. Add 5 ml of 40 per cent sodium hydroxide solution.
- ☆ Heat the distillation apparatus over the hot plate.
- ☆ Ammonia is liberated and collected in the boric acid.
- ☆ Distillation is continued till no more ammonia is coming, then disconnect the burner.

*3. **Titration***

- ☆ In the burette, take N/10 Hydrochloric acid. At first note the blank reading.
- ☆ Then the ammonia collected in the boric acid is titrated against the standard acid till the blue color is converted into pink color.

Observations

Blank	*Initial Reading (ml)*	*Final Reading (ml)*	*Vol. of N/10 HCl Used (ml)*
1			
2			

Let,

Vol. of N/10 HCl consumed is = V_3

Vol. of digested sample made = V_1

Vol. of aliquot taken = V_2

Weight of sample taken = W

Calculations

Per cent of Nitrogen

$$= \frac{0.0014 \times \text{Vol. of digested sample made} \times \text{Vol. of N/10 HCl used} \times 100}{\text{Wt. of sample taken} \times \text{Vol. of aliquot taken}}$$

$$= \frac{0.0014 \times V_1 \times V_3 \times 100}{W \times V_2}$$

Per cent crude protein = per cent Nitrogen × 6.25

Where, 6.25 is the general factor.

Method–IV

Protein (Dye Binding Method Corper, 1976)

Principle

The net increase in absorbance of Bromocresol Green at 630 nm on binding to serum albumin at 4.2 pH is directly proportionate to the concentration of albumin.

Reagents

1. *Buffered B.C.G dye solution*: Dissolve 8.85 g of succinic acid, 108 mg of B.C.G (sodium salt) and 100 g of sodium azide in about 950 ml of water, Add 4 ml of 30 per cent Brij-35. Adjust the pH to 4.2 and make volume to 1 litre with distilled water.
2. *Standard*: A pooled serum sample may be used (Standardized by Kjeldahl method) or human serum albumin of highest purity may be used. Bovine serum albumin is unsatisfactory.

Procedure

- ☆ Take 3 ml buffered dye solution in a series of tubes.
- ☆ Add 2 ml of serum sample and read immediately (within 30. sec) at 630 nm absorbance against dye solution.
- ☆ Analyze in duplicate and calculate the albumin content by comparison with standard run similarly

Note: Reading must be taken within 30 sec otherwise significant errors may result.

Method–V

(*i*) Estimation of Protein by Spectrophotometry (U.V. Method)

Principle

Light is one form of universal energy. The rays of light are nothing but forms of electromagnetic waves. These waves are having specific wavelength, frequency and energy level. When light passes through any light absorbing substances, there is absorption of energy. The degree of light absorption varies with source and nature of light as well as the concentration of and path length of the medium *i.e.* solution. Natural light consists of

1. Infra red rays (>750nm)
2. Visible rays (380-750nm)
3. Ultraviolet rays (<380nm)

If light is allowed to pass through a prism, the rays of different wavelengths get separated in the spectrum. A light of specific wavelength when passes through a solution containing light absorbing substances, the degree of absorption is related to its concentration and the length of light path and this can be explained with Beer's law.

Beer's Law

The absorption of light is directly proportional to the concentration of substance in a solution provided the length of light path and wavelength of light are constant.

According to Beer's law

$$A \propto C$$

where,

A = Absorbance

C= Concentration of solution

Reagents and Materials

1. Bovine serum albumin (BSA)
2. Distilled water
3. Spectrophotometer
4. Test tubes with stands
5. Beakers

Procedure

☆ Take six test tubes and serially number them from 1 to 5 for taking standard solution and one test tube no. 6 is used for blank reagent.

- ☆ The standard protein used is Bovine Serum Albumin BSA (1mg/ml)
- ☆ In test tubes number 1 to 5, protein is taken quantitatively *i.e.* 200 mg, 400 mg, 600 mg, 800 mg, and 1000 mg. This is done by taking BSA 0.2 ml, 0.4 ml, 0.6 ml, 0.8 ml and 1ml in the test tube 1, 2, 3 and 4, respectively.
- ☆ In the test tubes, distilled water was added in serial order quantitatively in the amount of 2.8 ml, 2.6 ml, 2.2 ml and 2.0 ml, respectively to make up the volume 3 ml in each test tube.
- ☆ In the test tube No. 6, only 3 ml distilled water was taken which was treated as blank.
- ☆ The contents of each tube were mixed thoroughly and O. D. or absorbance was recorded at 280nm.
- ☆ Taking OD on y-axis and concentration in x-axis, a standard curve was plotted in a graph paper.

Observations

Sample No.	*OD 280 nm*	*OD 260 nm*
(A)	–	–
(B)	–	–
(C)	–	–

Calculations

$$\text{Protein (mg/ml)} = 1.45 \times OD_{280nm} - 0.74 \times OD_{260\,nm}$$

N.B:

When protein is in very small quantity, the concentration of protein is derived by 1.44 (OD at 215 nm–OD at 225 nm) = mg/ml

Determination of Absorption Maxima (λmax) for Standard Protein Solution (Bovine Serum Albumin (BSA)

Principle

The relation between the transmittancy and absorbance of a solution containing light absorbing material and wave length of light passing through the solution is given by the so called absorption spectrum. It is estimated by quantitatively by measuring the transmittancy for a particular concentration and depth of solution at various wavelengths and plotting them. λmax is obtained which is the maximum absorption at a particular wavelengths of the spectrum.

Reagents and Materials

1. Standard Bovine Serum Albumin BSA (1mg/ml), 1ml BSA+2ml DW = 3 ml
2. UV Visible Spectrophotometer

Procedure

- ✰ Two test tubes are taken.
- ✰ One test tube contained 3 ml distilled water and the other contained 1ml BSA+2ml distilled water, total being 3 ml.
- ✰ The former is treated as blank and the latter as sample whose lmax is to be estimated.
- ✰ The spectrophotometer is made on for few min for stabilization. Then using the blank sample 0 and 100 per cent transmittancy was adjusted.
- ✰ Then the protein solution is kept in the cuvette and its absorbance or O.D. is recorded at different wavelength *viz.* 200, 220, 240, 260, 280, 300, 320, 340 and 360 nm. After getting the OD at various wavelengths, the result is plotted in a group and the λmax is determined.

Results

λmax of the standard protein solution is 280 nm.

(*ii*) Estimation of Protein by UV Method

Principle

The aromatic amino acids tyrosine, tryptophan, phenylalanine present in protein absorb maximum at 280 nm and sulphur constraining amino acids cystein, cystine and methionine and nucleic acids, present in the protein absorbs maximum at 260nm. This property is useful in estimating protein by UV method. Almost all proteins contain tyrosine residues.

Procedure

- ☆ Standard BSA (1mg/ml) was taken at two concentration of 50µg/ml and 100µg/ml in test tube "A" and "B", respectively.
- ☆ Buffalo serum (10 times diluted) was used after making final dilution of 100 times and taken in test tube "C".
- ☆ O.D. is recorded both at 280nm and 260nm for each sample in test tubes A,B and C and protein content is calculated.

Method–VI

Protein Estimation by Lowry's Method

Principle

This method combines the use of Biuret reaction of protein with Cu^{2+} ions in alkali and reduction of phosphomolybdic and phosphotungstic acid in Folin's reagent by the aromatic amino acids viz tyrosine, tryptophan, phenylalanine present in the protein to produce a more intense colored product which can be quantitatively estimated by a spectrophotometer at 750nm.

The method has the advantage of applicability to dried material as well as the solutions and it is very sensitive. Samples containing as little as 5µg of protein can be readily analyzed. The basis of the assay is thought to be the production of cuprous ions which reduce the Folin's reagent.

Reagents and Materials

1. *Copper Reagent*

   Composition

   1 per cent Copper Sulphate solution: 0.5 ml

   2 per cent Potassium sodium tartrate: 0.5 ml

   2 per cent Na_2CO_3 in 0.1N NaOH: 50 ml

   All three to be mixed properly

2. *Folin's reagent*: 1:1 diluted with distilled water (2N solution) or dilute 1 volume of commercial Folin's reagent with 1 volume of 0.1N NaOH.

3. *Protein standard solution* Bovine serum albumin (1mg/ml)

Procedure

- ☆ In the test tubes numbered 1, 2, 3, 4, 5 and 6 standard protein solution 10 times diluted (*i.e.* 0.1mg/ml) was taken 0.25, 0.5, 0.75, 1.0, 1.5 and 2.0 ml to get a concentration of protein 25μg, 50μg, 75μg, 100μg, 150μg, and 200μg, respectively was taken and in test tube No. 7, blank was taken and in the test tube No.8, 0.1 ml sample (unknown) was taken.
- ☆ The volume of each test tube was made up to 2 ml by using 0.1N NaOH. Then 2 ml of copper reagent was added to each test tube and waited for 10 min.
- ☆ Then 0.5 ml of Folin's reagent (1:1 dilution) was mixed into each test tube and waited fro 30 min. O. D. was recorded at 750nm.
- ☆ A standard curve was plotted using the protein concentration (μg) of standard solutions in X-axis and respective O.D. in Y-axis.
- ☆ Using this standard curve, the concentration of protein of unknown sample was estimated.

Method–VII

Protein (Modified Lowry's Procedure)

Principle

The final color is the result of (a) biuret reaction of protein with copper ion in alkali and (b) reduction of phosphomolybdic-phosphotungstic reagent by the tyrosine and tryptophan present in the treated protein.

Reagents and Materials

1. *Alkaline copper reagent*: Add 10 ml of 1 per cent potassium tartrate and 10 ml 0.5 per cent $CuSO_4$ mixing to 20 ml of 10 per cent NaOH. Then add 50 ml of 20 per cent sodium carbonate. Mix and make upto 100 ml with water.
2. *Phenol Reagent*: Dilute 1 volume of Folin-Ciacalteu's reagent (2N) with 16 volumes of distilled water just before use.
3. *Protein Standard*: Dilute bovine serum albumin stack solution with 1 per cent sodium dodecyl sulfate to give a

concentration of 100 µg/ml. Use 10-50 µg for the standard curve.

4. Spectrophotometer
5. Hot water bath
6. Test tubes and beakers

Procedure

- ☆ To 0.5 ml of sample blank and standard solutions taken in duplicate add 0.5 ml of alkaline copper reagent, mix and let it stand for 10 min.
- ☆ Then add 2.0 ml phenol reagent quickly to each tube.
- ☆ Mix immediately and heat in a water bath at 55°C for 5 min.
- ☆ Cool immediately in running water and read absorbance of the sample and standard at 650 nm against blank.
- ☆ Calculate the protein contents.

Calculations

$$\text{Protein (g/ml)} = \frac{\text{Reading of test sample}}{\text{Reading of standard}} \times \text{Conc. of standard}$$

Method–VIII

Protein Estimation by Biuret Method (Colorimetry Method)

Principle

Biuret reagent reacts with the peptide bonds of protein. There is formation of coordination complex between peptide bond of protein and cuprous ions of copper sulphate present in the Biuret reagent. This complex is blue in color and its absorbance is recorded at 540nm.

Reagents and Materials

1. Test tubes
2. Pipettes
3. Beakers
4. Reagent bottle

5. Chemical balance
6. Spectrophotometer

Biuret reagent- Composition

KI	2.5g
$CuSO_4$	1.5g
Potassium sodium tartrate	4.25g
NaOH	4g
Distilled Water	Upto 1 litre

Procedure

- ☆ Five test tubes were serially numbered from 1 to 5 and another test tube No. 6 was taken for blank reagent. The test tube No. 1 to 5 were used for standard solution of different concentration.
- ☆ The standard protein solution used was Bovine Serum Albumin, BSA (1mg/ml) *i.e.* which was consisting of 1000 µg protein per ml of solution.
- ☆ Protein of 200, 400, 600, 800 and 1000 µg *i.e.* 0.2, 0.4, 0.6, 0.8 and 1ml of BSA was taken in test tube No. 1, 2, 3, 4 and 5, respectively.
- ☆ Then volume of each test tube was made upto 1ml by adding 0.8, 0.6, 0.4, 0.2 and 0.0 ml distilled water to the test tubes in serial order.
- ☆ 2 ml Biuret reagent was added to each test tube to make up the volume to 3ml and mixed thoroughly using Cyclo mixer.
- ☆ In test tube No.6 only distilled water 1ml and Biuret reagent 2ml was taken to make up the volume to 3ml.
- ☆ Blue color was developed after 30 min. and OD was recorded at 540 nm
- ☆ Standard curve was plotted on a graph paper.
- ☆ Using standard curve, determine the protein percentage of the test sample.

3. Fat Estimation

Method–I

Extractable Fat (Soxhlet Method)

Application

The method is generally applicable but less precise than *i.e.* Weilgul, Rose, and Gottlief etc.

Principle

The fat is extracted with petroleum ether from the dried sample. The solvent is removed by evaporation and the residue of fat is weighed.

Reagents and Materials

1. Petroleum ether (40–60°C)
2. Soxhlet extraction apparatus
3. Air oven
4. Extraction Thimbles.

Procedure

- ✰ Take about 5g well ground dried sample in an extraction thimble.
- ✰ Place the thimble in the extractor and connect a weighed flask containing 100 ml petroleum ether. Connect the extractor to a reflux condenser.
- ✰ Extract the sample under reflux for 5-6 hours.
- ✰ Evaporate the petroleum ether extract to dryness and add 2 ml acetone. Blow air gently into the flask to remove the last traces of solvent.
- ✰ Dry the flask containing the fat residue in an air oven at 100°C for 5 minutes. Cool in a desiccator and weight.

Calculations

W_1= Net weight (g) of sample before drying

W_2= Weight (g) of flask empty

W_3= Weight (g) of flask with fat

$$\text{Extractable Fat (per cent)} = \frac{W_3 - W_2}{W_1} \times 100$$

Method–II

Total Fat (Chloroform-Methanol Extraction)

Application

The method is applicable all types of food products when further characterization of fat is required.

Principle

The fat is extracted by stirring the sample vigorously with chloroform methanol mixture at room temperature. A calculated amount of water is added to separate out two phases, the lower chloroform layer containing the fat. The chloroform layer is separated off, washed with dilute sodium chloride solution to remove extracted proteinaceous material and dried with anhydrous sodium sulphate. The chloroform extract is then evaporated to dryness and the fat residue is determined.

Reagents and Materials

1. Chloroform
2. Methanol
3. Magnesium chloride 20 per cent in H_2O
4. Sodium chloride 0.1 per cent in H_2O.
5. Sodium sulphate anhydrous
6. Sintered crucible No.1
7. Centrifuge
8. Desiccator

Procedure

- ☆ Weigh 5g of the sample in a 25 × 200 mm test tube.
- ☆ Add some glass beads, 5 ml of chloroform and 10 ml of methanol.
- ☆ Add 0.05 ml of 20 per cent $MgCl_2$ solution and mix on a whirl mixer for 2 minutes.
- ☆ Add more of 5 ml chloroform and mix again for 2 minutes.

- ✰ Add distilled water to bring the total water content (including that of the sample) to 9 ml.
- ✰ Mix for additional 4 minutes, and filter through sintered crucible porosity No. 1 into a 150 mm tube.
- ✰ Wash the residue and crucible with 3 × 2.5 ml chloroform.
- ✰ Swirl the tube contents and then centrifuge at 1500 r.p.m. for 5 minutes.
- ✰ Remove the upper aqueous layer.
- ✰ Add 10 ml of 1 per cent Sodium chloride to the chloroform extract and mix by gentle inversion for a number of times.
- ✰ Centrifuge at 1500 r.p.m. for 5 minutes.
- ✰ Remove the top aqueous layer as before.
- ✰ Add 1-2 g anhydrous powdered sodium sulphate. Stopper the tube tightly and shake vigorously and filter.
- ✰ Wash the tube crucible with 3 × 2.5 ml chloroform.
- ✰ Transform the chloroform extract to a pre-weighed vial.
- ✰ Place the vial on steam bath and evaporate of the chloroform.
- ✰ Finally place the vial in an oven at 100°C for 5 minutes.
- ✰ Cool in desiccator and reweigh.

Calculations

W_1 = Let weight (g) of sample

W_2 = Weight (g) of vial empty

W_3 = Weight (g) of vial + fat

$$\text{The Fat (per cent)} = \frac{W_3 - W_2}{W_1} \times 100$$

Note: Filter paper should not be used since phospholipids are absorbed on to the paper.

Method–III

Fat by Van Gulik Butyrometer

Sample is digested with an acetic perchloric acid mixture in boiling water and the centrifuged fat is measured in Van Gulik

Butyrometer, which is specially designed for use with 5 g of meat product The sensitivity of the readings is increased by containing much of the fat within the bulb, so that the graduated scale covers only the desired range of readings.

Reagents and Materials

1. Salwin acid
2. Van Gulik Butyrometer (Range 0-35 per cent)
3. Water bath with shaker
4. Gerber centrifuge.

Procedure

- ✩ Weigh 5 g meat product in a perforated glass cup mounted on a rubber stopper 'A'.
- ✩ Insert the above stopper into the dry Van Gulik Butyrometer (Figure 2.1).
- ✩ Place the butyrometer in a tall stand and add 12-13 ml of salwin acid reagent slowly through the narrow end so that the level is below zero mark.
- ✩ Place the tube in boiling water for about 15-30 minutes.
- ✩ Shake the tube thoroughly from time to time after inserting the small stopper 'D' until the sample is thoroughly dispersed and digested.
- ✩ Remove the stopper 'D' during heating.
- ✩ Insert the small stopper after digestion is over and shake the butyrometer.
- ✩ Transfer the butyrometer to a water bath (65-70°C) and keep it for 5 min.
- ✩ Centrifuge the butyrometer at 1100 rpm.
- ✩ Return the butyrometer to water bath (65-70°C) for 4 min. and carefully add more salwin reagent through the narrow end of the butyrometer so that the top of the lower layer is slightly below the zero mark.
- ✩ Return the butyrometer to water bath for 2 min.
- ✩ Centrifuge for 5 min.
- ✩ Return to water bath for 2 min.

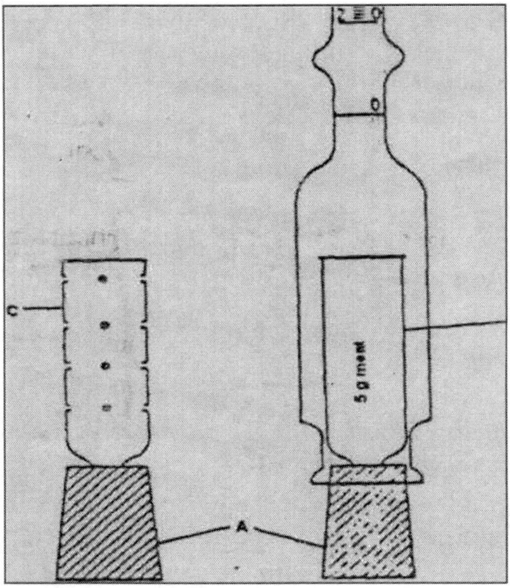

Figure 2.1: Van Gulik Meat Butyrometer for the Volumetric Estimation of Fat

☆ Read the fat percentage directly from the graduated scale at the bottom of the meniscus after adjusting the lower level of fat column to the zero mark by carefully manipulating the large stopper 'A'.

Method–IV

Fat Content by Mojonnier method

The fat content of a processed food product was determined by Mojonnier method with some modifications as per procedure given by ISI (1961). Accurately 10 g of the prepared sample was weighed into the tube. Added one ml of concentrated ammonia solution and mixed well in the lower bulb. Added 10 ml of the alcohol and mixed by allowing the liquid to flow backwards and forwards between the two bulbs. The tube was allowed to cool in cold running water. Added 25 ml of diethyl ether, closed with a bark cork and shaken vigorously for one minute.

Opened the tube and added 25 ml of light petroleum ether (40-60°C), closed the tube and shaken vigorously for one minute. The tube was allowed to stand on the flat bottom of the lower bulb until the ethereal layer was clear and completely separated form the aqueous layer.

Decanted carefully as much as possible of the supernatant layer into a suitable flask by gradually bringing the cylindrical bulb of the tube into a horizontal position. Extraction was repeated by using 15 ml of diethyl ether and 15 ml of petroleum ether.

Evaporated carefully the solvents from the flask and residual fat was dried in the oven at 100±1°C for one hour until successive weighings did not show a loss in weigh by more than one milligram. The fat in per cent was calculated as

$$\text{Fat (per cent)} = \frac{\text{Wt. of fat}}{\text{Wt. of sample taken}} \times 100$$

Method–V

Determination of Crude Fat (AOAC, 1975)

Reagent and Materials

1. Soxhlet extraction apparatus
2. Heating mantle
3. 50 ml beaker
4. Steam bath
5. Extraction thimble
6. Petroleum ether (40-60°C or 60-80°C)

Procedure

- ☆ Weight 3 to 4 g by difference into thimble containing small amount of sand or asbestos.
- ☆ Mix with glass rod, place thimble and rod in 50 ml beaker and dry in oven for 6 hours at 100-102°C or 1.5 hours at 125°C.
- ☆ Place the thimble in a Soxhlet extractor and connect it with the flask.

- ☆ Pour sufficient quantity of Petroleum ether to run the syphon.
- ☆ Place the extractor along with the flask on the heating mantle and connect the condenser with water tap.
- ☆ Switch on the heating element and as soon as ether begins to boil, adjust the heat so that the ether boils smoothly.
- ☆ Extract for 8 hours. After this period remove the thimble and collect the ether in the extractor so that only small quantity is left in the flask.
- ☆ Stop boiling and collect the ether from extractor in the bottle.
- ☆ Transfer the residual ether containing fat into tared beaker and wash the flask with redistilled ether in order to transfer last traces of the fat.
- ☆ Evaporate on steam bath in the fume cupboard.
- ☆ Place beakers in oven at 105°C for 30 minutes, cool in dessicator and weigh.

Calculations

$$\text{Per cent Ether extract} = \frac{W_2 - W_1}{W_0} \times 100$$

where,

W_1 = Weight of empty beaker

W_2 = Weight of beaker + ether extract

W_0 = Weight of sample

Method–VI

Estimation of Fat Content (Kramlich *et al.*, 1973)

Principle

Determination of the fat content of the moisture free sample is done by extracting the fat with a suitable solvent.

Reagents and Materials

1. 250 ml Skellysolve F (Technical grade solvent, chiefly hexane or petroleum ether).

2. Spatula
3. Soxhlet apparatus with heat controlled unit
4. Drying oven
5. Glass funnel
6. Analytical balance.

Procedure

- ☆ Transfer carefully the moisture free sample to an extraction thimble.
- ☆ Small particle of solids and separated fat in the aluminum can are removed by repeated washing with the solvent.
- ☆ Place extraction thimble in the extractor with an attached receiving flask and pour the solvent washer into the thimble through a glass funnel.
- ☆ Connect the extractor and receiving flask to the Soxhlet condenser.
- ☆ Adjust the electrical heating unit so that solvent syphons over 5 to 6 times per hour and extract the fat on the Soxhlet apparatus for 16 to 20 hour.
- ☆ Remove the extraction thimble and place it in the original aluminum can.
- ☆ Evaporate the remaining trace solvent from the fat free residue by drying in the 212°F oven (about 2 hour).
- ☆ With the aid of a spatula, carefully transfer all residue from the thimble to the aluminium can.
- ☆ Weigh the can containing the fat free residue.

Observations

W_1 = Weight of original sample

W_2 = Weight of empty can

W_3 = Weight of can and dried sample extract

W_4 = Weight of can with fat free residue

Calculations

$$\text{The Fat (per cent)} = \frac{W_3 - W_4}{W_1} \times 100$$

$$\text{Fat free residue (per cent)} = \frac{W_4 - W_2}{W_1} \times 100$$

Test Limitations

- Skellysolve F (hexane) in addition to the fat, may, on rare occasions, extract other materials present in the sample.

Precautions

- Skellysolve F is highly flammable. Use adequate ventilation and avoid open flames.

Method–VII

Estimation of Fat by Cold Percolation Method

The extraction of fat from the samples involves the use of non-polar *viz.* hexane, petroleum ether (40–60°C) which percolates through a column containing finely dried powdered sample along with anhydrous sodium sulphate. The oil gets eluted by the solvent and is collected in a conical flask.

Reagents and Materials

1. Anhydrous sodium sulphate
2. CCl_4/Hexane/petroleum ether (40–60°C).
3. Percolator
4. Glass beads
5. Glass wool
6. Glass flask
7. Pestle and mortar.

Procedure

- Take 5-10g of completely dried meat sample.
- Grind it to a fine powder in a pestle and morter and mix it well with 20-30 g of anhydrous sodium sulphate.
- Plug the percolator by putting cotton glass wool at the bottom of it. Put approximately 2 g of sodium sulphate (anhydrous) over the plug.

- ✰ Pack the powdered material slowly inside the percolator carefully. Ensure that whole of the material is packed properly, pack little cotton over the material and cover with a layer of anhydrous sodium sulphate.
- ✰ Add 10 ml solvent in the mortar and pestle to ensure complete removal of the material and transfer it into the percolator.
- ✰ Keep a flask below the percolator and fill the percolator with the solvent.
- ✰ Let the solvent tickle down under gravity. The solvent will flow down extracting fat from the material.
- ✰ Add more solvent to fill the percolator and allow it to pass through the column. Repeat this three times.
- ✰ Flash evaporate or distill off the solvent using water bath. When approximately 5 ml of solvent is left in the flask, remove it and transfer it into a weighed conical flask and keep it in the oven till the solvent is evaporated.
- ✰ Weigh to a constant weight after keeping in desiccator for few minutes.

  Calculate the fat/oil per cent by a following formula:

$$\text{Oil/Fat per cent} = \frac{X}{W} \times 100$$

Here

X = Weight of flask + fat

W = Weight of sample

Precautions

- ✰ Put off flame in the lab while extracting oil using organic solvents.
- ✰ The sample should be crushed finely along with anhydrous sodium sulphate.
- ✰ Column should not be packed too tight or too light.
- ✰ Use glass beads in the flask while evaporating the solvent to check bumping.
- ✰ Add some anhydrous sulphate to the extractant to remove last traces of moisture and then filter it off.

Method–VIII

Total Fat Content (ISI Hand Book, SP : 18 Part XII–1984 Page–13)

Principle

Boiling of the test portion with dilute Hydrochloric acid to free the occluded and bound lipid fractions, filtration of the resulting mass, drying of the fat retained on the filter and extraction with n-hexane or light petroleum.

Reagents and Materials:

1. Blue litmus paper.
2. Boiling chips.
3. Hydrochloric acid–4N solution, dilute 100 ml of concentrated hydrochloric acid (Sp. gr. 1.19) with 200 ml of water and mix.
4. n- Hexane or, alternatively, Light Petroleum.–distilling between 40°C and 60°C, and having bromine value less than 1. For either solvent, the residue on complete evaporation should not exceed 0.002 g/100 ml.
5. Analytical balance
6. Continuous or semi-continuous extraction apparatus–Soxhlet type with an extraction flash of about 150 ml.
7. Clock glass or Petri dish–diameter not less than 8 cm.
8. Cotton Flask–300 ml
9. Cotton Wool–defatted
10. Desiccators–containing an efficient desiccant.
11. Electrically heated drying oven–adjusted to operate at 103 ±2°C.
12. Electrically heated sand bath or water bath or similar suitable apparatus.
13. Extraction thimble.
14. Meat mincer.

Procedure

Preparation of sample : as described earlier.

- ☆ Weigh 3 to 5 g of the minced sample to the nearest 0.001 g into the 300 ml conical flask.
- ☆ Dry the flask of the extraction apparatus, provided with some boiling chips for one hour at 103±2°C in the drying oven, allow the flask to cool to room temperature in the desiccators and weigh to the nearest 0.001g.
- ☆ Add to the test portion 50 ml of the hydrochloride acid and cover the conical flask with a clock glass.
- ☆ Heat the conical glass on asbestos wire gauze by means of a gas burner until the contents begin to boil and continue boiling with a small flame for one hour and shake occasionally.
- ☆ Add 150 ml of hot water. Moisten a fluted filter paper held in a glass funnel with water and pour the hot contents from the flask on to the filter.
- ☆ Wash the flask and the cork glass thoroughly three times with hot water and dry in the oven.
- ☆ Wash the filter with hot water until the washings do not affect the color of the blue litmus paper.
- ☆ Put the filter paper on the clock glass or petri dish and dry for one hour in the oven at 103±2°C.
- ☆ Allow to cool. Roll up the filter paper and insert it into the extraction thimble.
- ☆ Remove any traces of fat from the clock glass or petri dish, using cotton wool moistened with the extraction solvent, and also transfer the cotton wool to the thimble.
- ☆ Place the thimble in the extraction apparatus. The paper should be handled either with tongs that can be rinsed or with paper cover slips on the fingers.
- ☆ Pour the extraction solvent into the dried flask of the extraction apparatus.
- ☆ Wash the inside of the conical flask used for the disintegration with hydrochloric acid and the covering clock glass with a portion of the extraction solvent and add it to the extraction flask.
- ☆ The total solvent quantities should be one-and-a-half to two times the capacity of the extraction tube of the apparatus.

- ☆ Fit the flask to the extraction apparatus. Heat the extraction flask for 4 hours on the heated sand bath or water bath.
- ☆ After extraction, take the flask containing the liquid from the extraction apparatus and distill off the solvent using the heated sand bath or water bath.
- ☆ Evaporate the last traces of the solvent on the water bath, using air blowing, if desired.
- ☆ Dry the extraction flask for 1 hour in the drying oven at 103±2°C and after allowing to cool to the room temperature in the desiccators, weigh to the nearest 0.001g.
- ☆ Repeat this process until the results of two successive weighings do not differ by more than 0.1 per cent of the sample weighed.
- ☆ Verify the completion of the extraction by taking a second extraction flask and extracting for a further period of 1 hour with a fresh portion of the solvent.
- ☆ The increase in weight should not exceed 0.1 per cent of the weight of the sample. Carry out two determinations on the same prepared sample.

Observations

M_2 = Mass in g, of the flask with the dried fat

M_1 = Mass in g, of the empty extraction flask with boiling chips

M_0 = Mass in g, of the test portion

Calculations

The total fat content of the sample, per cent by weight, is equal to:

$$\frac{M_2 - M_1}{M_0} \times 100$$

Comments

- ☆ Take the result as the average of the two determinations.
- ☆ The difference between the results of two determinations carried out simultaneously or in rapid succession by the same analyst should not be greater than 0.5 g of total fat per 100g of sample.

Method–IX

Free Fat Content: ISI Handbook (Sp : 18 Part XII)–1984. p. 14)

Principle

Extraction by means of n-hexane or light petroleum of the dried residue, removal of the solvent by evaporation, drying and weighing of the extract.

Reagents and Materials

1. Boiling chips
2. *n*-hexane or light petroleum*
3. Analytical balance
4. Soxhlet apparatus (Continuous/semi-continuous)
5. Cotton wool–defatted.
6. Desiccators.
7. Drying oven at103±2°C.
8. Hot water bath
9. Extraction thimble.
10. Meat mincer.

* *n*-hexane or light petroleum-distilling between 40-60°C and having a bromine value less than 1, the residue on complete evaporation should not exceed 0.002 g/100 ml.

Procedure

Preparation of sample: same as discussed earlier.

- ☆ Dry the flask of the extraction apparatus containing boiling chips for 1 hr at 103±2°C in the drying oven, allow the glass to cool to room temperature in the desiccators and weigh to the nearest 0.001g.
- ☆ Transfer the dried sample, quantitatively from the dish to the extraction thimble.
- ☆ Remove the last traces of the dried sample from the dish, using cotton wool moistened with the extraction solvent and transfer this cotton wool also to the thimble.

- ☆ Place the thimble in the extraction tube of the apparatus.
- ☆ Pour the extraction solvent into the flask of the extraction apparatus; the amount of solvent should be at least one and a half to two times the capacity of the extraction tube.
- ☆ Fit the flask to the extraction apparatus and heat the flask for several hours on the heated sand or water bath, according to the extraction rate and the apparatus used.
- ☆ A period of 6 to 8 hours is generally sufficient for extraction. After extraction, take the flask containing the liquid from the extraction apparatus and distill off the solvent using, for example, the heated sand or water bath.
- ☆ Evaporate the last traces of the solvent on the water bath, using air blowing if desired.
- ☆ Dry the flask for 1 hour in the drying oven regulated at 103±2°C and after allowing it to cool to room temperature in the desiccators weigh to the nearest 0.001 g.
- ☆ Repeat this process of heating and cooling until the results of two successive weighings do not differ by more than 0.1 per cent of the sample weight.
- ☆ Verify the completion of the extraction by taking a second extraction flask and extracting for a further period of 1 hour with a fresh portion of the solvent.
- ☆ The increase in mass should not exceed 0.1 per cent of the sample weight. Carry out two determinations on the same prepared sample.

Observations

M_2 = Mass in g, of the flask with the fat after drying.

M_1 = Mass in g, of the empty flask with boiling chips.

M_0 = Mass in g, of the sample taken for drying.

Calculations

The free fat content of the sample as a percentage by weight, is equal to

$$\frac{M_2 - M_1}{M_0} \times 100$$

Comments

Take the result as the average of the two determinations. Report the result to one decimal place. The difference between the results of two determinations carried out simultaneously or in rapid succession by the same analyst should not be greater than 0.5 g of free fat per 100 g of the sample.

4. Ash Estimation

Analysis of Ash Contents of Foods

The term "ash" refers to the inorganic residue remaining after total incineration of organic matter. The ash content is determined from the loss of weight, which occurs from complete oxidation of the sample at a high temperature (usually 500-600°C) through combustion and volatilization of organic materials. The ash obtained does not typically contain exactly the same type of minerals as those present in the original food because they are typically converted into a different molecular structure, such as oxides, phosphates, and sulfates. Some minerals may also be lost due to volatilization.

To ash a sample completely, heating continues until the resulting ash is uniform in color, most commonly white or gray and free from particles of unburned organic material. Ashing may be accomplished by incineration over an open flame, in a muffle furnace, in a closed system in the presence of oxygen, by wet oxidation in the presence of sulfuric, nitric, and perchloric acids alone or in mixtures, or by microwave digestion.

The analysis of ash is important for several reasons. The ash content can be regarded as a general indicator of product quality. Ash analysis can also be used to determine food adulteration. Abnormally high ash content suggests the presence of an inorganic adulterant, such as sand or dirt, which could be further evaluated by determining acid-insoluble ash. Total ash content may be used to indicate the nutritional value of food products. More precise nutritional information is obtained by performing elemental analysis on the resulting ash.

(*i*) Methods of Ashing

The two primary ashing procedures are dry ashing and wet ashing, which are eventually utilized to determine the mineral content of the food. In dry ashing, the organic matter of the sample is oxidized by complete incineration at a high temperature in the presence of oxygen, whereas, in wet ashing the sample is oxidized with a mixture of concentrated strong acids. A more recent development in ashing methodologies is the use of microwaves to increase the speed of sample digestion. In addition to these direct methods, there are indirect techniques for determining specific ash components in foods, such as conductometric methods for measuring their total electrolyte contents. The selection of an ashing methodology depends on the purpose for which the ash is prepared, the minerals to be analyzed, and the method of analysis to be used.

1. Dry Ashing

Dry ashing is the most standard method for determining the ash content of a food sample. The sample is commonly ignited at 550-600°C to oxidize all organic materials without flaming. The inorganic residue, which does not volatilize at that temperature, is called ash. Detailed dry ashing methodologies for specific food products can be found in the Official Methods of Analysis of AOAC International.

Materials

1. Muffle furnace
2. Porcelain crucibles with lids
3. Analytical balance
4. Desiccator
5. Tongs
6. Marking ink

Additional Equipment Required for some Samples

1. Bunsen burner
2. Hot-plate (thermostatically controlled)
3. Tripod, iron
4. Steam bath
5. Atmospheric oven
6. Wash bottle
7. Double deionized water

Procedure

- ✰ Place clean and labeled crucibles in a muffle furnace at 600°C for 1 h. Turn off the furnace and transfer partially cooled crucibles into a desiccator and cool to room temperature.
- ✰ Weigh crucibles quickly to prevent possible moisture absorption. Use metal tongs to move the crucibles after they are heated or dried (crucible preparation).
- ✰ Weigh accurately 2 g of dry food or 10 g of wet sample into the prepared pre-weighed crucible.
- ✰ For wet samples, dry the sample on a hot-plate or steam bath or in an atmospheric oven at 100°C for 1 h. Omit this step for dry sample.
- ✰ For samples under the hood. Place the crucible on an iron tripod over a Bunsen burner or hot-plate and slowly char for about 30 min. Heat cautiously to prevent excessive foaming and spattering. A few drops of olive oil can be added to reduce spattering. Omit this step for samples not rich in carbohydrates.
- ✰ Place the crucibles in muffle furnace set at 550°C for 12-18 h. Burn off the organic material until the samples become completely free from carbon and appear as light gray or white ash.
- ✰ If black carbon spots persist in the crucible due to incomplete oxidation of the samples, turn off the furnace, allow the sample to cool, add a few drops of concentrated nitric acid directly on to the unburnt spots, and switch on the furnace again. Raise the temperature to 600°C and leave 1-2 h depending on the incompleteness of the ashing.
- ✰ Turn off oven and allow it to cool. Opening a hot oven will cause ash to blow out of crucibles. Upon opening the cooled oven, cover crucibles with a lid or watch glass to prevent ash from blowing away.
- ✰ Transfer the partially cooled crucibles into a desiccator and continue cooling to room temperature. When cooled, weigh the crucible as quickly as possible to prevent moisture absorption.
- ✰ Save the ash sample if elemental analysis is to be performed.

Calculations

$$\text{Per cent Ash} = \frac{\text{Weight of residue}}{\text{Sample weight}} \times 100$$

Precautions

- ☆ Marking ink should be permanent type for crucibles as normal ink burns away.
- ☆ Moisture free samples should be taken for ash estimation.
- ☆ Charring of samples should be done to remove smoke before placing it into the muffle furnance

2. Wet Ashing

For samples that are to undergo elemental analysis, wet ashing is more advantageous than dry ashing, The wet ashing method is preferable to the dry ashing procedure for digesting organic matter prior to the determination of mineral constituents because it eliminates difficulties resulting from losses of more volatile inorganic constituents during dry ashing and slow dissolution of the residue after dry ashing.

Although using a single acid is desirable, it is often not practical for the complete digestion of the sample Nitric acid functions as a good oxidant, but typically volatilizes before the sample is completely oxidized. Sulfuric acid aids in the digestion, but using perchloric acid with nitric acid greatly facilitates decomposition of many hard-to-digest organic compounds. However, great care must be taken when using perchloric acid because it can be explosive. Thus, wet ashing commonly employs concentrated nitric and perchloric acids to oxidize the organic matter of the food sample. These acids are partially removed by volatilization and the soluble mineral constituents remain dissolved in nitric acid. Any silica present is dehydrated and made insoluble.

Reagents and Equipment

1. Nitric acid (69-71 per cent)
2. Perchloric acid (60-62 per cent)
3. Digestion acid*
4. 2 M Hydrochloric acid**

5. Fume hood
6. Hot-plate
7. Beaker
8. Volumetric flask
9. Watch glass
10. Goggles (protective glasses).
11. Glass rod
12. Glass beads
13. Hot air oven
14. Steam bath

* Digestion acid: add 1 volume perchloric acid to 4 volumes nitric acid.

** 2 M Hydrochloric acid: Add 1 volume hydrochloric acid (36 per cent HCl) to 5 volumes deionized water.

Procedure

- ☆ Weigh accurately 1-2 g dry, ground sample into a 250 mL beaker. While wearing safety goggles, add 20 ml digestion acid and three boiling beads to the beaker under a fume hood.
- ☆ After addition of the digestion acid, allow the reaction to proceed in the hood at room temperature for 3-4 h with occasional swirling using hands and a glass rod, or leave overnight.
- ☆ Place the sample beaker on a hot-plate maintained at a temperature such that the liquid in the beaker just simmers. When the initial reaction has subsided, slowly increase the temperature of the hot plate to 350°C.
- ☆ Evaporate the digestion acid until dense white fumes are formed within the beaker. Cover the beaker with a watch glass.
- ☆ Continue the digestion until the fluid in the beaker becomes clear and no visible particles remain. If the solution darkens when the volume is reduced, remove the flask from the hot-plate, add 1 or 2 ml nitric acid and continue the digestion with occasional swirling.

- ☆ Remove the beaker from the hot-plate. Rinse the watch glass with deionized, distilled water. Transfer the solution quantitatively to a 100 ml volumetric flask with deionized, distilled water.
- ☆ Stopper the flask, thoroughly mix the solution, and leave overnight. The water-insoluble and acid-insoluble silica will settle to the bottom and aliquots of the solution are drawn from the top, filtered, and stored in small polyethylene bottles for later analyses.
- ☆ The sample solution prepared can be used for the determination of various macro and trace minerals by atomic absorption spectrophotometry or other suitable method.

3. Soluble and Insoluble Ash

Ash samples are composed of soluble and insoluble portions in water and/or acids. To determine water insoluble ash, the residue in the crucible is solubilized with 10-25 ml deionized, distilled water after weighing the total ash. A watch glass is placed over the crucible to prevent loss by spattering, and it is heated just below boiling. The ash solution is filtered through ashless filter paper, and washed with several volumes of hot water. The filter paper and residue are dried, placed again in the crucible, ignited, and weighed. From that weight, the water-insoluble ash is determined, and the water-soluble ash is calculated by the difference from the value of the total ash content. The acid-insoluble ash can also be measured in a similar fashion to that of the water-insoluble ash. To determine acid-insoluble ash, water in the above procedure is replaced with 10 per cent hydrochloric acid (specific gravity 1.05) us the solvent. The acid insoluble ash typically represents contaminants, such as silicates in soil.

4. Microwave Digestion

Both dry and wet-ashing methods can be accelerated using microwave techniques. The major advantage is a significant time savings; however, the number of samples that can be ashed at a single time is reduced. Mineral analysis of shellfish after microwave digestion, dry ashing, and wet ashing indicated comparable results between the different methods, with dry ashing having consistently lower values than the two wet methods A comparison between mineral analysis results after various ashing techniques indicated that microwave digestion using nitric acid, hydrogen peroxide, and hydrogen fluoride yielded excellent recoveries of 13 minerals from several food types. Digestion using microwaves was generally found to be advantageous due to faster analysis times, improved recoveries, and lower values for blank analyses.

5. Conductometric Method

Conductometric methods are indirect methods for measuring the total electrolyte content of foods. Because foods rich in sugars are generally low in minerals, they require a 1arge amount of sample for dry or wet ashing. These samples also have a strong tendency to foam. Thus, conductometric procedures is a simple, rapid, and accurate method for estimating the ash content of these products. The basic principle of conductometric methods is that the mineral matter (*i.e.* ash) dissociates in solution while sucrose, being a nonelectrolyte, does not dissociate. Therefore, the conductance of the solution is an index of the ionic concentration, which in turn correlates with the mineral or ash content of the sample.

(*ii*) Methods of Ash Estimation

Method–I

Determination of Total Ash (AOAC, 1975)

Application

The method is applicable to all types of foodstuffs with the exception of high fat (7.50 per cent) foods.

Principle

Organic matter is burnt off at as low temperature as possible and the inorganic material is cooled and weighed. Heating is carried out in stages, first to driven of the water, then to char the product thoroughly and finally to ash at 550°C in a muffle furnace.

Procedure

1. Place the requisite number of silica crucibles in a muffle furnace to heat at 550°C for 15 minutes.
2. Remove the crucibles, cool in desiccators for one hour. Weigh each crucible.

3. Weigh accurately 5g of material into each crucible.
4. In case the sample is a liquid, pre-dry on a steam bath to prevent spitting during the charring stage.
5. Clear each sample on a hot plate till smoking ceases and the samples become thoroughly charred.
6. Place the crucibles inside the muffle furnace as near to the sample as possible and ash overnight at 550°C.
7. When the ash become clean and white in appearance, the crucibles should be removed from the furnace and be cooled in desiccators. If traces of carbon are still evident, cool the crucibles add a few ml of water and stir with a glass rod to break up the ash. Dry on a steam bath and then return to the muffle furnace for overnight.
8. When cool to room temperature, reweigh each crucible and ash.
9. By difference calculate the weight of ash.

Calculations

W_1 = Let weight (g) of sample

W_2 = Weight (g) of ash

$$\text{Then Ash value (per cent)} = \frac{W_2}{W_1} \times 100$$

Note

☆ If the ash is to be used for trace element or phosphorus determination, clean the crucibles by boiling in 6N HCl and rinse with distilled water.

☆ Too rapid heating is to be avoided since some of the salts will fuse and absorb carbon which is difficult to ignite. The use of too high and temperature may also cause some losses of volatile salts such as sodium and iron chloride.

Method–II

Estimation of Ash Content (Kramlich *et al.*, 1973)

Principle: Removal of organic material is done by heating and the remaining inorganic salts are determined gravimetrically.

Materials

1. Muffle furnace
2. Analytical balance
3. Porcelain crucibles
4. Pair of crucible tongs.
5. Spatula

Procedure

- ☆ Weigh an aliquot of fat-free residue into a porcelain crucible.
- ☆ Then place crucible in a muffle furnace not exceeding 500°C for 12 hours.
- ☆ Cool in desiccators. Then weigh ash and crucible.

Calculations

Weight of ash = Weight of ash and crucible – Weight of crucible

$$\text{Per cent ash} = \frac{\text{Weight of ash}}{\text{Weight of sample}} \times 100$$

Test Limitations

Ash content is dependent to a degree on the nature of the ash. Certain constituents (chlorides) may be volatilized, reduced (sulfates), or distilled as complexes during this early stage of ashing.

5. Carbohydrates Estimation

(*i*) Estimation of Crude Fibre

The carbohydrates of the food are contained in two fraction, the crude fibre (CF) and nitrogen free-extract (NFE). The CF fraction contains cellulose, lignin and hemi- cellulose, but not necessarily all of these materials present in the feed are in this (CF) fraction a variable proportion of them are contained in the NFE, depending upon the species and stage of growth of the plant material. The CF is a valuable fraction because of the correlation existing between it and the digestibility of the food, particularly in monogastrics. The bulkiness of food is due to CF and is important, since it gives satisfaction to the animals, and helps in the movement of digesta through the small intestine.

Method–I

Principle

The CF is the loss in weight on ignition of dried residue remaining after digestion/boiling of the fat free sample with 1.25 per cent of H_2SO_4 and 1.25 per cent of NaOH solution in turn for 30 minutes. This imitates the gastric and intestinal action in the process of digestion in monogastrics.

Reagents and Materials

1. 1.25 per cent w/v(0.225N) Sulphuric acid solution
2. 1.25 per cent w/v Sodium Hydroxide solution (0.313 N)
3. Iso-amyl alcohol
4. 600 ml spotless beaker (tall form)
5. Hot plate
6. Round bottom flask
7. Buckner's funnel
8. Muslin cloth
9. Wash bottle
10. Spatula
11. Clay crucible
12. Suction bottle
13. Suction pump
14. Desiccators
15. Pair of tongs
16. Muffle furnace
17. Analytical balance.

Procedure

- ☆ Weight about 2 g fat free sample from the thimble (if the fat content is less than 2.0 per cent the sample as such can be taken) and transfer the sample into the 600ml spoutless beaker.
- ☆ Add 200ml of 1.25 per cent H_2SO_4 to each beaker; place it on a hot plate and with heat regulator. Cover the beaker with a cold water filled round bottom flask. Add about

one ml of iso-methyl alcohol or dekalin (antifoaming agents), bring the contents to boil (takes 1-2 minutes) and continue boiling for another 30 minutes.

- ☆ After boiling for 30 minutes remove the beaker from hot filter the contents through a muslin cloth, fitted on top of a Buckner's funnel under suction. Make the residue acid free by repeated washing with hot water (test with litmus paper).
- ☆ Transfer quantitatively the residue left on the muslin cloth to the same beaker with the help of spatula.
- ☆ Add 200ml of 1.25 per cent of NaOH solution into the beaker and boil for exactly 30 minutes, while covering the beaker with cold water flask. Filter again through the muslin cloth and make the residue free of alkali by repeated hot water washing (test with litmus paper)
- ☆ Transfer the residue thus obtained quantitatively into a previously weighed and numbered clay crucible with the help of 10-15 ml of distilled water. Dry the crucibles containing residue at 100°C to a constant weight (over night). Cool and record the weight again.
- ☆ Ignite the residue in a muffle furnace at about 550°C for 3-4 hours.
- ☆ Cool the crucibles in the desiccator and record its weight. CF then can be determined by the loss in weight due to ignition.

Observations

W = Weight of sample (g)

W_1 = Weight of crucible + residual material after oven drying (g)

W_2 = Weight of crucible + residual material after ashing(g)

Calculations

$$\text{Per cent of CF} = \frac{\text{(Weight of crucible + dried residue before ashing)} - \text{(Weight of crucible + its content after ashing)}}{\text{Weight of sample}} \times 100$$

$$\frac{W_1 - W_2}{W} \times 100$$

Precautions

- ☆ Swirl the beaker occasionally during boiling to bring down the feed material sticking to the walls of the beaker.
- ☆ To maintain the acidity/alkalinity of the digestion media during boiling cover the beaker with round bottom flask containing cold water and change it frequently when it becomes hot.
- ☆ Transfer of residual material from the muslin cloth should be done quantitatively. Make use of spatula and wash bottle for this purpose.

Observations	*R1*	*R2*
Weight of sample		
Weight of crucible + residual material after oven drying (W_1)		
Weight of crucible + residual material after ashing (W_2)		

Calculations

Per cent CF =

Method–II

The method is applicable to all food stuffs including meat and meat products.

Principle

A fat free sample is treated with boiling sulphuric acid subsequently boiling sodium hydroxide. The residue after subtraction of the ash is regarded as fibre.

Reagents and Materials

1. Hydrochloric acid 1 per cent v/v

2. Sulphuric acid stock solution[1]
3. Sulphuric acid working solution [2]
4. Sodium hydroxide stock solution[3]
5. Sodium hydroxide working solution [4]
6. Antifoam[5]
7. Crucibles
8. Desiccators
9. Boiling water bath
10. Beaker (1lt)
11. Crude fiber apparatus
12. Weighing balance

1 Sulphuric acid stock solution 10 per cent w/v (Take 55 ml conc. H_2SO_4 and dilute to 1 litre).

2 Sulphuric acid working solution (Dilute 125 ml of the stock solution to 1 litre).

3 Sodium hydroxide stock solution 10 per cent w/v. (Dissolve 100gm NaOH in water to dilute 1 litre.)

4 Sodium hydroxide working solution 1.25 per cent. (Dilute 125 ml of the stock solution to 1 litre.)

5 Antifoam: 2 per cent Silicon antifoam in CCl_4.

Procedure

- ☆ Weigh (1-2g) of fat free dried sample in a 1 litre beaker.
- ☆ Add 200 ml 1.25 per cent (hot) H_2SO_4 and few drops of antifoam.
- ☆ Heat to boiling within 1 minute on the crude fibre apparatus.
- ☆ Keep the solution boiling exactly for 30 minutes under bulb condensers. Beaker may be rotated occasionally to mix the contents and remove the particles from the side.
- ☆ Filter the contents of the beaker through Bunchner funnel.
- ☆ Wash the sample back into the tall beaker with 200 ml 1.25 per cent sodium hydroxide. Brought to boiling point.
- ☆ Boil for exactly 30 minutes.

- ✰ Transfer all insoluble matter to the sintered crucible by means of boiling water till acid free.
- ✰ Wash twice with alcohol.
- ✰ Wash three times with acetone.
- ✰ Dry at 100°C to constant weight.
- ✰ Ash in a muffle furnace at 550°C for 1 hour.
- ✰ Cool crucible in a desiccators and reweigh.

Calculations

W_1 = Let weight (g) of sample

W_2 = Weight (g) of insoluble matter

W_3 = Weight (g) of Ash

$$\text{Crude fibre (per cent)} = \frac{W_2 - W_3}{W_1} \times 100$$

(*ii*) Determination of Neutral Detergent Fiber (NDF)

Apparatus

1. Refluxing apparatus
   - (*a*) Tall pyrex or corning beakers (spoutless) of about 500 ml capacity
   - (*b*) Round bottom flask as condenser
2. Sintered glass crucibles with coarse porosity (Grade 1) of about 50 ml capacity
3. Electronic balance
4. Vacuum pump
5. Hot plate
6. Wash bottle
7. Hot air oven and muffle furnace

Reagents

1. Neutral detergent solution (NDS):

   Distilled water : 1 litre

   Sodium lauryl sulphate: 30g

Disodium ethylene diamino tetra acetate (EDTA) dehydrate : 18.61g

Sodium borate decahydrate : 6.81g

Disodium hydrogen phosphate, (anhydrous): : 4.56g

2-ethoxyethanol (ethylene glycol monoethyl ether) : 10ml

Put EDTA and $Na_2B4O7.10H_2O$ together in a large beaker, add some of the distilled water and heat until dissolved, then add to solution containing disodium lauryl sulphate and 2-ethoxy ethanol. Put

Na2HPO4 in a beaker, add some of the distilled water and heat until dissolved, then add to solution containing other ingredients. Check pH range 6.9 to 7.1. If solution is properly made pH adjustment is rarely required.

2. Decahydronaphthalene (Decalin)- Reagent grade
3. Acetone- Use grade that is free from color and which leaves no residues upon and evaporation
4. Sodium sulphite (anhydrous)

Procedure

- ☆ Take 0.5 to 1.0g dry sample ground to pass 20 to 30 mesh (1 mm) into a beaker of the refluxing apparatus.
- ☆ Add in order 100 ml (preheated) NDS, 2 ml of decalin and 0.5 g sodium sulphite with a calibrated scoop and reflux for 60 minutes, time starting from the onset of boiling.
- ☆ Filter off the reagent, wash thrice with hot with hot distilled water under vacuum, remove vacuum, break up mat and wash crucible with hot water.
- ☆ Wash twice with acetone in same manner and suck dry. Dry crucible at 100°C for 8 hrs or overnight and weigh it.
- ☆ Report yield of recovered NDF as percent of cell wall constituents, estimate cell soluble material by subtracting this value from 100.
- ☆ Ash residues in the crucible for 3 hrs at 500-550°C and weigh. Report ash content as ash insoluble in neutral detergent.

Observations

Empty wt. of crucible = —— g

Wt. of dry sample = —— g

Wt of crucible + cell wall constituent = —— g

Wt of crucible + Ash = —— g

Calculation

1. Cell wall constituent (per cent) (NDF) =

$$\frac{(\text{Wt of crucible} + \text{cell wall constituents}) - \text{Wt. of crucible}}{\text{Wt. of dry sample}} \times 100$$

2. Cell contents (per cent) = 100-cell wall constituents

3. Insoluble ash in neutral detergent (per cent) =

$$\frac{\text{Wt of crucible} + \text{Ash} - \text{Wt of crucible}}{\text{Wt. of dry sample}} \times 100$$

(*iii*) Determination of Acid Detergent Fibre (ADF)

Principle

The acid detergent fibre procedure provides a rapid method for lignocellulose determination in feeds stuffs. The residue also includes silica. The difference between cell walls and ADF is an estimate of hemicellulose, however, this difference does include some protein attached to cell walls. The acid detergent fibre is used as a preparatory step for lignin determination.

Apparatus

Same as used in NDF estimation.

Reagents

1. Sulphuric acid (H_2SO_4) Reagent grade, standardize to 1N (100 per cent assay) is 49.04g dissolved in 1000 ml.
2. Cetyl trimethyl ammonium bromide (CTAB) Technical grade-20g. Weigh sulphuric acid and make up to volume with distilled water. Check normality by titration before addition of detergent. Then add CTAB and stir.

3. Decalin (Reagent Grade)
4. Acetone. Use grade that is free from color and leaves no residue upon evaporation.
5. *n*-Hexane (Technical Grade)

Procedure

- ✰ Weigh 1 g air dry sample ground to pass 20 to 30 mesh (1mm) screen or approximate equivalent of wet material in a beaker suitable for refluxing
- ✰ Add 100 ml cold (room temp.) acid detergent solution and 2 ml decalin. Heat to boiling in 5 to 10 minutes. Reduce heat as boiling begins, avoid foaming. Reflux 60 min from on set of boiling; adjust boiling to a slow level.
- ✰ Filter on a previously weighed crucible. Wash with hot distilled water 3-4 times breaking the mat. Repeat wash with acetone twice or until it removes no more color and suck dry.
- ✰ Optional wash with hexane. Hexane should be added while crucible still contains some acetone (Hexane can be omitted if lumping is not a problem in lignin analysis). Suck the acid detergent fibre free of hexane and dry at 100°C for 8 hrs or overnight and weigh after cooling of crucible in desiccator.

Observations

Empty wt. of crucible = —— g

Wt. of dry sample = —— g

Wt of crucible + fibre = —— g

Calculation

Acid detergent fibre per cent on dry matter basis=

$$\frac{(\text{Wt of crucible + fibre}) - \text{Empty weight of crucible}}{\text{Wt. of dry sample}} \times 100$$

(*iv*) Estimation of Glucose and Glycogen (Raghuramulu *et al.,* 1983e)

Principle

Glycogen, a polysaccharide is found in the liver, muscle, kidneys and other tissues but is notably absent from brain. It is the chief storage carbohydrate of muscle and the available ready source of substrate for rapid glycolsis in activity. During extremely active muscular work skeletal muscle uses its glycogen as energy source via glycolsis. During recovery some of the lactate formed in the muscles is transported to the liver and rebuilt to form blood glucose, which returns to the muscles to replenish their glycogen store.

On hydrolysis with amylase, it yields maltose and with acid it yields glucose. It thus resembles starch and dextrin in these respects but differ significantly in molecular architecture. In the glycogen molecule the component glucose residues are linked in chains by a-1, 4 and 1, 6 glycoside bonds to give a highly branched structure.

Glycogen is soluble in cold water to form an opalescent solution and ordinarily gives a red color with iodine, although some forms of glycogen which give blue or purple color with iodine are known. These color differences are apparently related to the extent of chain branching, the blue color representing relatively unbranched chains and the red color to the highly branched chains.

In the following method of glycogen estimation, glycogen present in the meat tissue, extracted and saturated Na_2SO_4 precipitate it when mixed with ethanol. It is reprecipitated by 95 per cent ethanol. On acid hydrolysis by HCl or H_2SO_4, reducing sugar D-glucose is liberated which is determined by means of Anthrone Method. Red color obtained is estimated in a colorimeter at 620 nm to the concentration of D-glucose.

Prevent Biochemical Degradation

Original muscle composition is preserved as much as possible by rapid cooling. These muscles can be rapidly cooled by quick immersion into isopentane, cooled in liquid nitrogen or Freon cooled in liquid air. Thick tissue pieces are rapidly compressed between two cold blocks of metal; this is conveniently done with specially designed pliers cooled in liquid air.

After freezing, the tissue is powdered at liquid nitrogen temperature. This can be done in a cooled mortar or by high speed shaking for about one min in a stainless cartridge with a steel ball at liquid nitrogen temperature.

Reagents and Materials

1. 30 per cent KOH
2. Saturated Na_2SO_4
3. 95 per cent ethanol
4. 0.6M or 5 M HCl or H_2SO_4
5. 0.5M NaOH (Phenol red)
6. Glucose standard solution*
7. Anthrone Reagent**.

* Glucose standard solution containing 111 μg or 0.5 μM (90μg)

** Anthrone Reagent: dissolve 0.2g of anthrone in 100 ml of 95 per cent H_2SO_4. This should be prepared fresh or kept in the refrigerator for not more than 2 days.

Procedure

- ☆ 100 mg of powdered tissue is extracted by boiling in 3ml of 30 per cent KOH for 20 to 30 min. or until the tissue has dissolved.
- ☆ The glycogen is precipitated by adding 0.5 ml of saturated Na_2SO_4 and 3.5 ml of 95 per cent ethanol heating again until the mixture boils.
- ☆ After cooling, the mixture is centrifuged in the cold at 3000 rpm and the solution discarded.
- ☆ The sediment is dissolved in 2 ml water and re-precipitated with 2-5 ml of 95 per cent ethanol.
- ☆ The collected sediment is hydrolyzed for 2 to 2.5 h in 6 ml of 0.6 M HCl or H_2SO_4 or for 30 min. in 2 ml of 5 M acid, in a boiling water bath.
- ☆ The hydrolyzate is cooled, neutralized with 0.5 M NaOH (Phenol red) and made upto a volume dependent on the expected glycogen content generally 50-100ml.

- ☆ For determination by the Anthrone method, 5ml of neutralized hydrolyzate (containing 15 to 150 μg of glucose) is measured into a test tube.
- ☆ A second tube receives 5ml of glucose standard containing 111 μg or 0.5 μM (90 μg), a third one contains 5 ml water to serve as a blank.
- ☆ Immerse the tubes in cold water and add 10 ml of anthrone reagent.
- ☆ The tubes are covered with glass marbles and heated for 10 min. in boiling water, cooled immediately and read in a colorimeter at 620 nm.
- ☆ In calculating the glycogen content of the tissue, keep in mind that 1 g of glycogen yields 1.11 g of glucose on hydrolysis. The glucose standard can be made up to allow for this.

Digestion of Starch by Salivary Amylase

Principle

The most important storage polysaccharides in nature are starch, characteristic of plant cells and glycogen in animal cells. Both starch and glycogen occur intracellularly in the form of large crystals or granules. Starch when extracted from granules with hot water, will form turbid colloidal solution or dispersions. Starch is especially abundant in tubers such as potatoes and in seed *e.g.* corn and wheat, rice etc.

Starch contains two types of glucose polymer, α-amylase and amylopectin. The former contains D-glucose units connected by α-1, 4 linkages. The glycosidic linkages joining successive glucose residues in amylopectin are a-1, 6 linkages.

Salivary amylases are in the form of α-amylases which hydrolyze the α-1, 4 linkages of the outer branches of glycogen and amylopectin to yield D-glucose, some amount of maltose and a resistant "core" called a limit dextrin. Salivary amylase cannot attack the α-1, 6 linkages at the branch points.

Reagents and Materials

1. Buffer solution*
2. Rice starch

3. Water bath
4. 1N acetic acid
5. Conical flask 250 ml
6. Test tubes, pipettes etc
7. 3, 5-Dinitrosalicylate reagent**
8. Standard maltose solution***
9. Spectrophotometer

* Sodium chloride 0.2g, KH_2PO_4 1.515g and Na_2HPO_4: $2H_2O$ 1.98g are dissolved in distilled water, made volume 500 ml and pH adjusted to 6.9 to prepare buffer solution

** 3, 5-dinitrosalicylate reagent: Dissolve 10g of 3, 5-dinitrosalicylic acid, 300g of Rochelle salt and 16g of NaOH in CO_2 free water and dilute to 1000 ml Store in brown flasks, protect from CO_2.

*** Standard maltose solution:Dissolve 200 mg of maltose monohydrate in distilled water and dilute to 100 ml. As a preservative 0.27g of benzoic acid may be added.

Procedure

Preparation of Starch Solution

- ☆ Weigh out 2g of rice starch and make a suspension of starch in 100 ml of cold buffer.
  - ☆ Boil on water bath for 15 min under careful stirring

Digestion of Starch

- ☆ In a flask A, mix 0.1 ml of undiluted saliva with 50 ml of distilled water
- ☆ Into another flask B, add 50 ml of distilled water
- ☆ Place two flasks in constant temperature water at 37°C.
- ☆ Place two other flasks C and D, each containing 50 ml of starch in the same water bath.
- ☆ After about 10 min, pour the contents of flask C into flask A, mix and immediately transfer 10 ml of mix solution to a test tube containing 1 N acetic acid, to interrupt the reaction.
- ☆ Take new samples after 1,2,3,5,10,30,60 and 120 min and treat them similarly.

- ☆ In the same way mix the contents of the flasks D into flask B and withdraw samples at similar intervals.
- ☆ The samples obtained are analyzed for viscosity, reducing power and iodine color. It will be seen that the decrease in viscosity proceeds much more rapidly than does the increase in reducing power, which clearly shows that the salivary amylase is an endoamylase (α-amylase).

Viscosity

Measure Specific viscosity (viscosity/viscosity of water) of each sample in an Ostwald viscosimeter.

Reducing Power

The amount of reducing groups liberated is determined with the 3,5 dinitrosalicylate reagent, originally described by Sumner (1924) and Hostettler *et al.* (1951). In contrast to most other methods for reducing sugar determination this method does not require previous protein precipitation.

Procedure

- ☆ In a series of test tubes pipette 0.5 ml of the starch digested samples and 1.5 ml of water.
- ☆ For a blank pipette out into another tube 2 ml of water and for a standard series, prepare tubes with 0.5, 1.0, 1.5 and 2 ml of the standard maltose solution plus water to a total volume of 2ml.
- ☆ To each of the tubes add 2 ml of the 3, 5-dinitrosaliccylate reagent.
- ☆ Mix the contents and immerse the tubes in a boiling water bath for 10 min.
- ☆ Cool with running tap water for 2 min
- ☆ Add 20 ml of water to each tube mix and measure the red color produced in a spectrophotometer at 530 nm. using 1 cm cuvettes.

Iodine color

To 2ml of the sample add 0.1 ml of iodine solution (50 mg of iodine dissolved in 50 ml of 2 per cent potassium iodine solution) and 20 ml of water, mix and note the color.

Estimation of Glucose Content in Liver

Liver principally rich in glycogen content therefore first liver glycogen is hydrolysed to glucose and glucose thus formed is estimated by the following methods:

(*a*) Glucose Oxidase method

(*b*) Ferricyanide method

(*c*) Hagedorn and Jensen method

(*d*) Nelson and Somogyl method

Conversion of Liver Glycogen to Glucose

Reagents

1. 30 per cent KOH
2. 95 per cent Ethanol, 60 per cent ethanol
3. 2N H_2SO_4
4. Phenol Red Indicator

Procedure

☆ Take the liver immediately after slaughter and remove the excess blood by processing under the folds of filter paper.

☆ Put the sample into a weighed stoppered test tube containing 30 per cent KOH and weigh it again

☆ Adjust the volume of alkali by 2 ml per g of liver tissue.

☆ Cool the test tube in ice cold water.

☆ Then add two volumes of 95 per cent ethanol and mix it vigorously.

☆ Boil the content but avoid spurting

☆ Leave the content overnight at the refrigerated temperature.

☆ Centrifuge it at 5000-7000 rpm for 3-5 min.

☆ Take the precipitate and dissolve in warm water, again centrifuged and re-precipitated with 2 volumes of 95 per cent ethanol.

☆ The precipitate is centrifuged again and washed 6-10 times with 60 per cent ethanol.

☆ Two ml of 2 N H_2SO_4 per g of liver is added and hydrolysed in a boiling water for 3-4 hr.

- ✩ The solution is neutralized with NaOH using phenol red as indicator. Filter it again.
- ✩ Glucose is determined in an aliquot by above mentioned methods.
- ✩ The factor 0.93 is used to convert glucose to glycogen

Method–I

Glucose Oxidase Method (Raghuramulu *et al.*, 1983a)

Principle

Glucose is oxidised to gluconic acid by glucose oxidase. The hydrogen peroxide liberated is indicated by peroxidase and the oxygen transferred to an acceptor, which is colorless in the reduced form colored in the oxidised form.

Reagents

1. *Protein precipitant:*

   Sodium tungstate, $Na_2WO_4 2H_2O$: 10 g

   Disodium phosphate Na_2HPO_4: 10 g

   Sodium chloride, NaCl: 9 g

   Dissolve in about 800 ml water and add approximately 125 ml of 1 N HCl to adjust to pH 3.0. Add 1 g of phenol and make up to 1L with water. Stable for 1 year at 25°C.

2. *Color reagent:*

   Sodium azide: 0.3 g

   4-aminophenazone: 0.1 g

   Disodium phosphate: 3.0 g

   Dissolve in 295 ml of water, then add 5 ml Fermoozyme 952 DM. Stable for 8 weeks at 4°C.

3. *Standard*: Prepare a saturated solution of benzoic acid, by dissolving 1 g of benzoic acid in about 300 ml of hot distilled water. Filter whilst hot, through a coarse filter paper. Needle like crystals should form on cooling. Immediately before use, filter to remove crystals and use the solution for making up the standards.

The stock standard contains exactly 1 g of pure glucose in 100 ml benzoic acid solution. The working standard is made by diluting

1 ml of stock standard with 49 ml with benzoic acid solution. Stable at 25°C for 1 year.

Procedure

- ☆ Pipette 0.1 ml of aliquot into 2.9 ml or protein precipitant.
- ☆ Mix well and centrifuge for about 5 min.
- ☆ A standard curve is set up for each batch of determination.
- ☆ To clean tubes pipette 0.1, 0.2 and 0.3 ml of the working glucose standards equivalent to 60, 120 and 180 mg per 100 ml. In each case make up the volume to 1 ml with protein precipitant reagent.
- ☆ Similarly, take a 1 ml of clear supernatant from the test and place in clean tubes.
- ☆ For the reagent blank use 1 ml of protein precipitant.
- ☆ To all tubes add 3 ml of color reagent and incubate at 37°C for 10 min.
- ☆ Then place the tubes in cold water for 1 min and read the absorbance at 505 nm against the reagent blank without further delay.
- ☆ Plot the absorbance of the standards graphically and read off the glucose values of the tests from this. If the glucose concentration of the test is greater than 180 mg/100 ml repeat the color development stage using a smaller aliquot of the supernatant, *e.g.* 0.2 or 0.5 ml. Include additional standards in the calibration where necessary and multiply the result obtained by the appropriate factor, depending on the volume of supernatant used.

Note

- ☆ Fermoozyme 952 DM (Hughes and Hughes Ltd.) contains both glucose oxidase and peroxidase.
- ☆ The chromogen, 4-aminophenazone replaces O-tolidine and dianisidine, which are thought to be carcinogenic.

Method-II

Ferricyanide method (Raghuramulu *et al.*, 1983b)

Principle

Sugar is oxidised with alkaline potassium ferricyanide and the ferrocyanide produced is measured photometrically after the conversion.

Reagents

1. *Ferricyanide solution*: Dissolve 0.5 g potassium ferricyanide in 1L of distilled water. Stop in brown bottle.
2. *Carbonate-cyanide reagent*: Dissolve 5.3 g sodium carbonate and 0.65 g potassium cyanide in 1L of distilled water.
3. *Ferric ammonium sulfate solution*: 1.5 g ferric ammonium sulfate in 1L of 0.05 N H_2SO_4.
4. *Tungstate solution*: 10 per cent sodium tungstate.
5. *Standard glucose solution*: 90 mg glucose in 100 ml of saturated benzoic acid solution.
6. N/12 H_2SO_4.

Procedure

- ☆ Take 0.02 ml of aliquot and add 0.2 ml of N/12 H_2SO_4.
- ☆ Add 8.78 ml of distilled water and mixed thoroughly.
- ☆ Add Sodium tungstate 0.02 ml and the solution mixed with a glass rod to precipitate the proteins.
- ☆ This is allowed to stand for 10 min and centrifuged. One ml of the above supernatant is mixed with 1 ml of ferricyanide solution and 1 ml of cyanide-carbonate solution.
- ☆ The solution is heated in a boiling water bath for 15 min and cooled immediately under running tap water.
- ☆ Add 5 ml of ferric ammonium sulfate (60 ml ferric ammonium sulfate + 0.12 ml H_2SO_4) immediately after cooling the solution.
- ☆ The solution is mixed and readings taken at 690 nm on a spectrophotometer after 10 min and a within 30 min.
- ☆ The standard glucose is diluted as per the range required and stated as above.

Calculations

A standard graph is drawn and the value for the serum sample is obtained from it (say X). This value multiplied by the dilution factor gives the glucose value in the sample.

0.02 ml blood is diluted to 9 ml than diluted to100 ml. Therefore, dilution factor is

$$\frac{100 \times 9}{0.02} = 45000$$

µg glucose/100 ml blood = 45000 × X

$$\text{mg glucose/100 ml blood} = \frac{45000}{1000} \times X = 45 \times X$$

Method–III

Hegedorn and Jenson Method (Raghuramulu *et al.*, 1983c)

Principle

The proteins present in liver aliquot are precipitated with zinc hydroxide. The reducing sugars in the protein-free filtrate reduce potassium ferricyanide on heating. The amount of unreduced ferricyanide is determined iodimetrically.

Reagents

1. *0.45 per cent zinc sulphate*: Prepared fresh every week by dilution of 45 per cent stock solution,
2. *0.1 N NaOH*: Prepared fresh every week by dilution of 2 N NaOH.
3. *Potassium ferricyanide solution*: 1.65 g recrystallised potassium ferricyanide and 10.6 g anhydrous sodium carbonate are dissolved in 1 L water and stored in a dark bottle, protected from light.
4. *Iodide-sulphate-chloride solution*: Zinc sulphate 10 g and NaCl 50 g are dissolved in 200 ml water. On the day of use 5 g potassium iodide is added for 200 ml.
5. 3 per cent Acetic acid
6. *0.005 N Sodium thiosulphate*: Prepare fresh daily by the dilution of 0.5 N sodium thiosulphate. 0.5 N sodium thiosulphate is prepared by dissolving 70 g of the salt in 500 ml water. It is better if a solution of a slightly higher

normality is prepared, since sodium thiosulphate decomposes soon. The solution should be protected from light and stored in the cold. The normality is checked daily with 0.005 N potassium iodate.

7. *0.005 N Potassium iodate*: This is a stable solution and should be made accurately. 0.3566 of the anhydrous salt is weighed accurately and dissolved in 2 L water. This is used to check sodium thiosulphate and potassium ferricyanide solutions.
8. *Starch indicator*: 1 g soluble starch is dissolved in 100 ml saturated NaCl solution.

Procedure

- ☆ One ml 0.1 N NaOH and 5 ml 0.45 per cent zinc sulphate are pipetted into a test tube.
- ☆ Take 0.1 ml of aliquot with micro-pipette and introduced into the gelatinous zinc hydroxide in the test tube rinsing out the pipette twice with the mixture.
- ☆ The tube is kept in a boiling water bath for 3 min and cooled without disturbing the precipitate.
- ☆ The mixture is filtered through a Whatman No. 42 filter paper of lightly pressed moistened cotton.
- ☆ The tube is washed twice with 3 ml portions of water and filter into the same containers.
- ☆ 2 ml potassium ferricyanide is then added to the filterate and heated in a boiling water bath for 15 min.
- ☆ After cooling 3 ml iodide-sulphate-chloride solution followed by 2 ml 3 per cent acetic acid are added.
- ☆ The liberated iodine is then titrated against 0.005 N sodium thiosulphate, using 2-3 drops of 1 per cent starch as indicator towards the end of the titration.
- ☆ A blank is run through the entire procedure simultaneously. The blank should give a titre value of 1.97 to 2.00 ml.

Calculations

The value may be expressed as mg glucose/100 ml by

consulting the table that is given in the Appendix. The value for blank is subtracted from the one for the unknown.

$$\text{mg glucose}/100\ \text{ml} = \frac{0.385}{2.0} \times X \times 1000$$

where,

X = 2.0-B (titre value of sample) in terms of exactly N/200 thiosulphate.

Method–IV

Nelson and Somogyl Method (Raghuramulu *et al.,* 1983d)

Principle

The liver aliquot is heated with alkaline copper reagent and the reduced copper formed is treated with arsenomolybdate reagent resulting in the formation of violet color which is read in the photometer.

Reagents

1. 5 per cent zinc sulphate solution.
2. 0.3 N Barium hydroxide.

   These two solutions should be so adjusted that 5 ml zinc sulphate require 4.7 to 4.8 ml barium hydroxide for complete neutralization.
3. Alkaline copper reagent:

   Solution A: 25 g anhydrous Na_2CO_3, 25 g Rochelle salt, 20 g Na_2HCO_3 and 200g anhydrous H_2SO_4 are dissolved in about 800 ml water and diluted to 1 L. The solution is stored at room temperature and never below 20°C. It is filtered before use if any sediment is formed.

   Solution B: 15 per cent $CuSO_4\,5H_2O$ containing one or two drops of concentrated H_2SO_4.

   On the day of use 25 parts of solution A and 1 part of B are mixed. This is the alkaline copper reagent.
4. *Arsenomolybdate color reagent*: Ammonium molybdate 25 g is dissolved in 450 ml water of ml concentrated H_2SO_4 is added and mixed. Disodium orthoarsenate ($Na_2H\,AsO_4 7H_2O$) 3.0 g dissolved in 25 ml water and is added

with stirring to the acidified molybdate solution. It is then reduced in an incubator at 37°C for 24–48 hr and stored in a glass stoppered brown colored bottle.

5. Standard glucose solutions (Stock glucose solution): Exactly 1.0 g of anhydrous pure glucose dissolved in 10-15 ml 0.2 per cent benzoic acid and diluted to 100 ml with benzoic acid solution.

Three working standards are prepared by diluting 0.5, 1.0 and 2.0 ml of the stock solution to 10 ml with benzoic acid. These solutions in benzoic acid keep indefinitely at room temperature.

Procedure

- ☆ 0.1 ml aliquot is introduced through a clean dry micropipette into a test tube containing 3.5 ml of water and mixed well.
- ☆ 0.2 ml of 0.3 N Barium hydroxide is added to the tube.
- ☆ After the mixture turns brown 0.2 ml Zinc sulphate is added and mixed. After 10-15 min the mixture is filtered through Whatman No. 1 filter paper.
- ☆ ml aliquot of the filtrate is transferred into 2 separate test tubes.
- ☆ One ml alkaline copper reagent is then added.
- ☆ The tubes are covered with a glass marble and placed in a boiling water bath for 10 min. The tubes are then cooled under running water.
- ☆ One ml arsenomolybdate reagent is added and the solution diluted to 25 ml with water.
- ☆ Simultaneously standard and a reagent blank are similarly prepared.
- ☆ The intensity of color produced is read at 500 nm. The color is stable and readings may be at convenience.

Calculations

$$\text{mg glucose/100 ml} = \frac{\text{Reading of unknown}}{\text{Reading of standard}} \times \text{Conc. of standard} \times 4 \times \frac{100}{0.1} \times \frac{1}{100}$$

6. Estimation of Calorific Value of Meat Samples Using Bomb calorimeter

Principle

A known quantity of a properly dried meat sample is ignited electrically and burnt in excess of oxygen in the bomb calorimeter. The maximum temperature rise is measured with the thermometers in a controlled system. By comparing this rise with that obtained when a sample of known calorific value is burnt, the calorific value of the sample material can be determined.

Reagents and Materials

1. Meat sample
2. Benzoic acid
3. Distilled water
4. Barium hydroxide
5. Sodium carbonate
6. Gallenkamp and Ballistic Bomb calorimeter
7. Steel crucible
8. Firing wire
9. Cotton thread

Procedure

- ✰ Weigh 0.5-1.0 g of finely ground representative meat sample and make a pellet with the help of a pellet press.
- ✰ Weigh the samples for dry matter determination at the time of pelleting.
- ✰ Put the pellet in a preweighed crucible and weigh again.
- ✰ Put the bomb top on the stand. Thread a piece of fuse wire through the electrodes and tie to it a single strand of cotton. Keep the lengths of fuse wire and cotton thread constant in order to facilitate the calculations of calorific value.
- ✰ Swing the crucible into position, clamp the ring and arrange the ends of cotton thread so that they are in contact with the sample.
- ✰ Pipette 1 ml of distilled water into the bomb.

- ☆ Place the electrode assembly into the bomb body ensuring that it fits correctly.
- ☆ Tighten the bomb closure ring by hand only.
- ☆ Fill the bomb to 25 atmospheric pressure with oxygen.
- ☆ Fill water into calorimeter vessel to submerge the bomb completely. The vessel and water should give a total weight of 3 kg. The quantity of water used is not critical but it must be constant for all tests to an accuracy of ± 0.5 g.
- ☆ Place the bomb on three supports in the calorimeter vessel and check for the gas leakage that the bomb should not show any sign of gas leakage.
- ☆ Gently slide the top of the calorimeter console on to the bomb. Switch on the main and press down the bomb firing plug to contact the bomb.
- ☆ Adjust the initial temperature and press the fire switch.
- ☆ After 8 minutes read temperature on main thermometer. Note the final temperature when it get stabilizes.
- ☆ Switch off the main switch. Remove the bomb from the vessel.
- ☆ Release pressure of the bomb using pressure release cap.
- ☆ Open the bomb and wash the electrodes and inside top and body of the bomb and distilled water.
- ☆ Collect these washings in a beaker for correction of nitrogen and sulphur contents. However, there is no need of corrections for N and S, when benzoic acid is used as standard.

Calculations

$$\text{Bomb equivalent} = \frac{(6318 \times W) + A}{T}$$

W = Weight of benzoic acid (g)

A = Correction factor for wire, thread, N and S (Heat of combustion of thread and wire may be taken as 3692 cal/g and 1400 cal/g (or 2.3 cal/cm) respectively.)

T = Rise in temperature

$$\text{Gross energy} = \frac{(\text{Bomb equivalent} \times \text{T}) - \text{A}}{\text{Dry weight of meat sample (g)}}$$

Estimation of Calorific Value (Energy)

When the samples are ignited and burnt in excess oxygen in the bomb, the rise in temperature is measure by the thermocouple and galvanometer system. This is compared by burning a standard sample of known calorific value and the energy value is determined. Gross energy of a meat product is determined by Gallenkamp Ballistic Bomb Calorimeter (Haque and Murari Lal, 1999)

Requirements

1. Bomb calorimeter
2. Galvanometer
3. Standard benzoic acid
4. Oxygen gas cylinder

Procedure

1. Take approx. 1 g of dried meat sample and weighed in the pre-weighed steel crucible.
2. This crucible is placed on the support pillar in the base of the bomb calorimeter.
3. The firing wire and sample are connected with the help of cotton thread.
4. The bomb is fired under an oxygen pressure of 25 atm.
5. The initial and final temperature readings on the galvanometer are noted.
6. The deflection on the galvanometer was compared with the deflection caused by 1 g of benzoic acid as standard with known calorific value (6.318kcal/g).
7. The calorific value of the sample is calculated and expressed as kcal/100g.

Chapter 3
Physico-chemical Properties

1. Meat Color

The color is one of the most important quality attributes both for raw materials and finished food products. To ensure color accuracy and uniformity of raw materials before they go into production line, is the responsibility of the food processor. The visible color of foods is due to coloring pigments like anthoacyannis, carotenoids, chlorophylls. Xanthophylls etc. or due to other pigments produced during the processing and storage. The quantitative determination of these pigments is carried out spectrophoto-metrically after extraction from samples, which is an elaborate, time consuming and costly technique.

1.1. Estimation of Total Meat Pigments (Myoglobin)

Method–I (Aragnosa and Henrickson, 1969)

Principle

Myoglobin (Mb) is constantly converted to MetMb and OxyMb. Here the myoglobin is converted to Cyanometmyoglobin in the presence of $K_3Fe\ (CN)_6$ and KCN, oxidizing agents, forming a stable compound. CyanoMetMb gives rise to orange/red color, which has maximum absorbance at 540 nm.

Reagents and Materials

1. Chilled distilled water
2. Potassium ferricyanide solution (50 mg/ml)
3. Potassium cyanide solution (10mg/ml)
4. Minced Meat
5. Pestle and Mortar
6. Refrigerator
7. Refrigerated centrifuge
8. Measuring cylinder
9. Spectrophotometer

Procedure

- ☆ Ten g of raw minced meat is taken in a mortar and 30 ml of chilled distilled water was added into it.
- ☆ It was homogenized or macerated thoroughly by pestle and the slurry was kept overnight at 4°C.
- ☆ The slurry was centrifuged at 5000 rpm for 10 min using refrigerated centrifuge.
- ☆ Supernatant was collected in 50 ml measuring cylinder.
- ☆ The residue was reextracted with 20 ml chilled distilled water.
- ☆ The supernatant was pooled and the volume was made upto 50 ml by adding distilled water.
- ☆ Take 20 ml aliquot, 0.1 ml $K_3Fe(CN)_6$ (50 mg/ml) and 0.1 ml KCN (10 mg/ml) were added.
- ☆ Optical density was recorded at 540 nm.

Calculations

$$\text{mg pigment/g tissue} = \frac{\text{Optical density} \times 17000\ (0.02 + d) \times 1000}{\text{Wt. of sample (g)} \times 11300}$$

$$= \frac{\text{O.D} \times 75.212}{\text{Weight of sample (g)}}$$

11300 = Molar extension coefficient of CyanoMetMb

17000 = Equivalent weight of meat pigment (myoglobin)

0.02 = Aliquot in litre

d = Volume increases d = 0.1 ml

1000 = to convert g to mg

Method–II (Warris, 1979)

Total Meat Pigments

Reagents and Materials

1. Chilled distilled water
2. Phosphate buffer (0.04 M, pH6.8)
3. Potassium ferricyanide solution
4. Sodium cyanide
5. Minced Meat
6. Pestle and Mortar
7. Refrigerator
8. Refrigerated centrifuge
9. Measuring cylinder
10. Spectrophotometer
11. Homogenizer

Procedure

- ✰ Samples (5g) of post rigor pig and sheep M. *Longissimus dorsi* which had been minced through 4.5 mm plate were extracted with 25 ml ice-cold 0.04 M phosphate buffer pH 6.8.
- ✰ They were homogenized for 20 sec using a Silverson laboratory homogenizer with microtubular head.
- ✰ After standing for 1 h at 4°C, the homogenates were centrifuged at 6500 g for 10 min and the residues reextracted twice.
- ✰ After adding a few micrograms of Potassium ferricyanide and sodium cyanide to convert the pigments to the CyanoMet forms, the supernatants were centrifuged at

30,000 g and 15° C for 1 h and the absorbance read at 540 nm.

- ☆ Total pigment concentration were calculated by multiplying the absorbance by 1.45, a factor derived from the extinction coefficient of cyanmet myoglobin given by Drabkin (1950) and a dilution factor based on the water content of meat taken as 75 per cent wet weight

Calculations

$$O.D. \times 1.45$$

N.B.

The need to clear extracts made with this buffer by high-speed centrifugation could be overcome by use of Hornsey's (1956) acetone-HCl extraction followed by determination of acid haematin

Method–III: Estimation of Total meat pigments

Total meat pigments are determined by solvent extraction technique modified from Hornsey (1956).

Reagents and Materials

1. Acetone
2. Distilled water
3. Concentrated HCl
4. Ultra Turrex Tissue homogenizer
5. Spectrophotometer

Procedure

- ☆ Take 10 g of the minced meat sample into a 100 ml beaker.
- ☆ The sample is homogenized to a smooth paste using an Ultra Turrax tissue homogenizer.
- ☆ Simultaneously add 23 ml of mixture containing 40 ml of acetone, 2 ml of distilled water and 1 ml of concentrated HCl.
- ☆ The remainder 20 ml of the solution is added and kept for 1 hour with intermittent mixing.
- ☆ Then the solution is filtered through Whatman's filter paper no. 42.

- ✩ This gives a solution of acid haematin in 80 per cent acetone (including the 7 ml water present in 10 g meat).
- ✩ The filtrate is composed of haematin derived from any uncombined meat pigments present, together with that resulting from the oxidized pigments.
- ✩ The optical density of this filtrate is measured at 640 nm, with an 80 per cent acetone/water solution as blank.
- ✩ The absorbance of the filtrate multiplied by a factor 680 gave the concentration of total pigments present in the meat as ppm of haematin.

Haeme iron contents are calculated as follows by using the haematin concentration.

Haematin (ppm) (a) : Abs (640 nm) × 680

Haeme iron: (a × 8.82)/100

1.2. Determination of Metmyoglobin Content in Meat

Principle

Metmyoglobin content of the given meat sample was estimated based on the principle of absorption maxima of myoglobin, oxymyoglobin and metmyoglobin at 525 nm, 572 nm and 700 nm wavelengths (Chu *et al.*, 1987; Krzywicki, 1979; Trout, 1989)

Reagents and Materials

1. Refrigerator
2. Meat homogenizer
3. Refrigerated centrifuge
4. Whatman filter paper No.42
5. Spectrophotometer
6. Phosphate buffer containing Sodium dihydrogen orthophosphate (NaH_2PO_4: $2H_2O$) and Disodium hydrogen orthophosphate or Sodium pyrophosphate dibasic: dihydrate (Na_2HPO_4: $2H_2O$).

Procedure

- ✩ Three gram of minced meat was taken and cold phosphate buffer 0.04 M pH 6.8 was added to the minced meat; Buffer: Meat = 10:1 ratio.

- ☆ The meat sample was homogenized for 20 sec.
- ☆ The homogenized sample was kept at 4°C for 1 h.
- ☆ Then it was centrifuged at 10,000 rpm for 5 min.
- ☆ The supernatant was filtered using Whatman filter paper No.42.
- ☆ Optical density (O.D.) was recorded at 525 nm, 572 nm and 700 nm.
- ☆ Percent Metmyoglobin was calculated according to the formula of Krzywicki, (1979).

Calculations

$$\text{Met Mb\%} = \left[1.395 - \frac{(OD_{572} - OD_{700})}{(OD_{525} - OD_{700})}\right] \times 100$$

Note

All heated samples were analyzed for PMD and percent metmyoglobin, pH was measured in all samples before heating.

1.3. Percent Mb Denatured (PMD) and Percent Metmyoglobin (PMM)

Reagents and Materials

1. Refrigerator
2. Meat homogenizer
3. Refrigerated centrifuge
4. Whatman filter paper No.42
5. Spectrophotometer
6. Phosphate buffer*

* Phosphate buffer: Sodium dihydrogen orthophosphate (NaH_2PO_4: $2H_2O$) and Disodium hydrogen orthophosphate or Sodium pyrophosphate dibasic: dihydrate (Na_2HPO_4: $2H_2O$).

Procedure

- ☆ Undenatured myoglobin was extracted from samples with cold (0°C) 0.04 M phosphate buffer pH 6.8 (Warris 1979).
- ☆ With heated samples, myoglobin was extracted by blending the complete sample with the phosphate buffer for 1 min on high speed using an Osterizer Imperial blender.

- ✰ With unheated samples, a 10g sample was homogenized with phosphate buffer for 20 sec on setting 4 using a homogenizer fitted with a point 10 probe generator.
- ✰ Sample to buffer ratio used for extraction of the different samples was as follows: heated pork and turkey, 1:4, heated beef, 1:5, and all unheated samples, 1:10
- ✰ Homogenates were centrifuged at 50,000 × g, 5°C for 30 min and the supernatant was filtered through a Whatman No. 1 filter paper.
- ✰ Absorbance of the filtrate was measured at 525, 572 and 700 nm using a spectrophotometer.
- ✰ Percent MMb and myoglobin concentration were calculated using the following formula (Krzywicki 1979)

$$\text{MMb\%} = \left[1.395 - \frac{(A_{572} - A_{700})}{(A_{525} - A_{700})}\right] \times 100$$

$$\text{Myoglobin (mg/ml)} = (A_{525} - A_{700}) \times 2.303 \times \text{dilution factor}$$

where,

$A\lambda$ = Absorbance at λnm

PMD was calculated using the following formula

$$\text{PMD} = 1 - \frac{\text{Myoglobin concentration after heating}}{\text{Myoglobin concentration before heating}} \times 100$$

1.4. Metmyoglobin/Myoglobin/Percent Mb Denatured Trout (1989) and Warris (1979) with Slight Modification

Reagents and Materials

1. Phosphate buffer
2. Homogenizer
3. Refrigerated centrifuge
4. Whatman filter paper No. 4

Procedure

- ☆ 3g of minced meat was taken.
- ☆ Cold Phospahte buffer 0.04M pH 6.8 was added to the minced meat @ buffer: meat = 10: 1 all unheated samples, 5: 1 heated beef, heated pork and turkey 4: 1.
- ☆ The meat sample was homogenized for 20 sec.
- ☆ The homogenized sample was kept at 4°C for 1 hr.
- ☆ Then it was centrifuged at 10,000 rpm for 5 min.
- ☆ The supernatant was filtered using Whatman filter paper No. 4.
- ☆ Optical Density (O.D.) was recorded in a spectrophotometer at 525, 572 and 700 nm.

Calculation (Krzywicki, K (1979)

$$\text{Metmyoglobin (\%)} = \{1.395\text{–}[(A_{572}\text{–}A_{700})/(A_{525}\text{–}A_{700})]\} \times 100$$

$$\text{Myoglobin (mg/ml)} = (A_{525}\text{–}A_{700}) \times 2.303 \times \text{dilution factor.}$$

$$\text{PMD} = \left[1 - \left(\frac{\text{Myoglobin conc. after heating}}{\text{(Myoglobin conc. before heating)}}\right)\right] \times 100$$

1.5. Meat Pigments in Cured Meat Products

Reagents and Materials

1. Acetone (Solution1 and 2)
2. Polypropylene centrifuge tubes with covers (45 ml).
3. Glass stirring rod with tapered end
4. Two Pyrex test tubes 15 × 90 mm
5. Spectrophotometer
6. Glass funnels–50 mm diameter
7. Watch glasses of 2 inches diameter.

Procedure

- ☆ Take 2 g of meat sample in a polypropylene centrifuge tube.
- ☆ Add 9 ml acetone (solution 1) to the centrifuge tube by help of a pipette.

- ☆ Macerate the meat thoroughly with the glass-stirring rod.
- ☆ Stopper the tube with centrifuge tube cover and mix by gentle swirling.
- ☆ Allow to stand for 10 minutes and then filter through Whatman filter paper No. 42 (9 cm diameter) into a test tube.
- ☆ The above operations should be carried out in a very subdued light to lessen fading of pigments during extraction.
- ☆ Transfer the filtrate into a 1 cm Beckman cell.
- ☆ Read the optical density within 1 h at 540 nm and calculate as nitroso pigment
- ☆ Prepare another 2 g sample as above and transfer to another polypropylene centrifuge tube.
- ☆ Add 9 ml acetone (solution 2) by help of a pipette.
- ☆ Macerate the meat sample thoroughly with the stirring rod.
- ☆ Allow to stand for 1 h before filtering.
- ☆ Filter the extract into test tube as above.
- ☆ Read the optical density within 1 h at 640 nm and calculate as total pigment.

Calculations

$$\text{Conversion (per cent)} = \frac{\text{ppm nitroso pigment}}{\text{ppm total pigment}} \times 100$$

where,

$$\text{ppm nitroso pigment} = \text{O.D. (540 nm)} \times 290$$

$$\text{ppm total pigment} = \text{O.D. (640 nm)} \times 680$$

Observations

Sample No.	*Optical Density*		*ppm nitroso pigment*	*ppm total pigment*	*Conversion (per cent)*
	at 540 nm	*at 640 nm*			
1.					
2.					
3.					

Precautions

- ☆ The procedure has been scaled down to a small sample size, therefore, the meat sample should be as homogenous as possible,
- ☆ Filtrate must be crystal clear.
- ☆ Percent conversion is calculated on a theoretically possible 100 percent of all hematin can actually be converted to nitroso pigment.

2. Muscle Fiber Diameter

Determination of muscle fiber diameter of skeletal muscle (Tuma *et al.*, 1962)

Reagents and Materials

1. 40 per cent formalin in physiological saline
2. Compound microscope containing caliberated micrometer
3. Blender

Procedure

- ☆ About 10 g of minced meat sample was fixed in 40 per cent formalin in physiological saline.
- ☆ The tissue was placed in an automatic mixture blender with enough formalin saline to cover blender blades. The blades were reversed to avoid cutting the fibers.
- ☆ The blender was run for 30 seconds at low speed to tear the fibers.
- ☆ Fifty straight fibers were measured using a compound microscope with 10 × object lens and a 10 × eye piece containing a calibrated micrometer.
- ☆ Muscle fiber diameter was measured at three places for each fiber and average value was calculated.
- ☆ Ocular micrometer 1 division = 13 μ

Observations

Sl.No.	*Division of Ocular Micrometer*				*Muscle Fiber Diameter (Avg. divisions × 13) μ*
	Upper	*Middle*	*Lower*	*Average*	
1.					
2.					

Metod–II

Estimation of Muscle Fibre Diameter

The fibre diameter of buffalo meat samples are assessed according to the method outlined by Jeremiah and Martin (1982).

Reagents and Materials

1. 0.25 M Sucrose solution
2. 1mM EDTA
3. Ultra Turrex Homogenizer
4. Light microscope

Procedure

- ☆ Take five grams of the minced meat sample
- ☆ Add 30 ml solution containing 0.25 M sucrose and 1 mM EDTA (ethylene diamine tetra acetic acid)
- ☆ Homogenize it in a Ultra Turrax tissue homogenizer at low speed for two 15s periods inter-spaced with a 5s resting interval to produce a slurry.
- ☆ One drop of the slurry is then transferred on to a glass slide and converted with a cover slip.
- ☆ The suspension is examined directly under a light microscope with 10X objective and 8X eyepiece equipped with calibrated micrometer.
- ☆ Muscle fiber diameter is measured as the mean diameter of the middle and the two extremities of the 25 randomly selected muscle fibers and expressed in micrometer.

3. Estimation of Sarcomere Length

Sarcomere length plays vital in measuring the textural quality of meat. It is determined by fixing the muscles according to Cross *et al.* (1980). The samples of sarcomere length measurements are removed after 24 hrs of ageing.

Reagents and Materials

1. 2.5 per cent gluteraldehyde
2. 5mM EDTA

3. Boric acid
4. Potassium chloride
5. Sucrose
6. Ultra Turrex Homogenizer
7. Micrometer
8. Light microscope

Procedure

- ☆ Prepare One gram of sample and put in 15 ml of first fixative or solution A (0.1 M KCl, 0.039 M boric acid and 5 m MEDTA in 2.5 per cent glutaraldehyde) in a 50 ml beaker.
- ☆ Cover it and keep it for 2 hours.
- ☆ Then the sample is transfer to 15 ml of storage solution or solution B (0.025 M KCl, 0.039 m boric acid and 5mM EDTA in 2.5 per cent glutaraldehyde) in a 50 ml beaker and again covers it and keep for 24 hours.
- ☆ The samples can be stored at 4±1°C until analysis.
- ☆ The fixed samples are quantitatively transferred to 50 ml polycarbonate centrifuge tubes.
- ☆ Add 7 ml of 0.25 M sucrose solution.
- ☆ Homogenize the sample is at a low speed with Ultra Turrax tissue homogenizer.
- ☆ A drop of the homogenate is transferred to a slide and covered with a cover slip.
- ☆ The slide is then examined under a microscope using a 10X eye piece with a calibrate micrometer under oil immersion objective lens (100X).
- ☆ Sarcomere length is measured as the mean length of 10 sarcomeres in 10 randomly selected myofibrils.

4. W-B Shear Force

Tenderness is an important meat quality trait that influences the consumer acceptability. Various factors influencing meat tenderness are type of animal, age, muscle location, connective tissue content, muscle fibre diameter etc. In animal body the tenderest muscle is *Psoas major* and most tough muscle is *Sternocephalicus.*

Tenderness of muscle can be measured both by resistance to mechanical shearing and by palatability test. Shear force value is an objective measurement of tenderness. In this, the force required to shear the muscle by mechanical means is calculated by using instrument called Warner-Bratzler Shear Press or Kramer Shear Press. In recent years, Texture Analyser or Texturometer is being used to measure the tenderness and different texture profiles of muscle.

The shear force can be estimated by Warner-Bratzler Shear Press and recorded (in kg) as per the method of Berry and Stiffler (1981) with slight modifications. Ten observations were recorded for each group of meat sample to get the average value.

Reagents and Materials

1. Meat sample
2. Water bath
3. Warner-Bratzler Shear Press
4. Sampling core
5. LDPE bags

Procedure

- ☆ Cook the meat samples at 80°C for 20 minutes in a water bath and cool in a refrigerator
- ☆ Prepare samples of 1cm^2 or round samples with a core sampler.
- ☆ Place the sample on cutting blade of W-B shear press (sample should be placed perpendicular to the fibre direction for shearing).
- ☆ Switch on the shearing machine *i.e.* Warner Bratzler Shear Press.
- ☆ Meat sample is cut slowly by the machine blade and the needle is deflected due to force of shearing the sample.
- ☆ Generally ten readings for each sample are taken and average value is calculated.
- ☆ This force is expressed as kg/cm.

5. Measurements of Myofibrillar Fragmentation Index (MFI)

Method–I (Method of Olson *et al.*, 1976 modified by Culler *et al.*, 1978)

MFI is related to the degradation of myofibrillar proteins during post mortem storage. The degradation mainly occurs due to CAF and endogenous neutral protease. It is useful to predict the level of tenderness in fresh and cooked steaks. It depend on the type of muscle and storage temperature. The measurements of MFI should be coincided with WB shear values and microscopic observations. It is more sensitive method than WB shear values and sensory evaluation. MFI is considered as more effective parameter for the measurement of tenderness of loin steak than collagen solubility or sarcomere length

Reagents and Materials

1. 100 mM Potassium Chloride (KCl)
2. 20 mM Potassium dihydrogen phosphate (KH_2PO_4)
3. 1 mM EDTA,
4. 1 mM Magnesium Chloride
5. Refrigerated centrifuge
6. Auto mixi blender
7. Spectrophotometer

Procedure

- ✰ 4g minced meat homogenized for 30 seconds in 40 ml of chilled isolating medium consisting of 100 mM KCl, 20 mM KH_2PO_4, 1 mM EDTA, 1 mM $MgCl_2$ centrifuged for 15 minutes.
- ✰ The sediment was re-suspended in 40 ml of isolating medium and centrifuged for 15 minutes and supernatants were discarded.
- ✰ The sediment was re-suspended in 10 ml isolating medium and passed through a polyethylene strainer (18 mesh) to remove connective tissue and debris.
- ✰ An additional 10 ml were used to facilitate passage of myofibrils through the strainer.

- ☆ Protein concentration of the suspension was determinant by Biuret method of Gormall *et al*. (1949).
- ☆ An aliquot of the myofibril suspension was diluted with isolating medium to obtain protein concentration of 0.5 ± 0.05 mg/ml.
- ☆ The diluted myofibril suspension was stirred and OD was read at 540 nm in a spectrophotometer.

Calculations

M.F. Index calculated by multiplying O.D. by 200.

Method–II

Estimation of Myofibrillar Fragmentation Index

The myofibrilla fragmentation index (MFI) is determined as described by Davis *et al*. (1980) with slight modifications. This basically measures the proportion of muscle fragments that passed through the muslin cloth after sample had been subjected to a high speed homogenization treatment.

Reagents and Materials

1. 0.25M sucrose solution
2. 0.02 M potassium chloride
3. Ultra Turrex Homogenizer
4. Centrifugation machine
5. Hot air oven/Incubator
6. Whatman Filter paper No. 1
7. Glassware

Procedure

- ☆ Take Ten grams of minced meat samples and transfer it to a 100 ml polycarbonate centrifuge tube Add 50 ml of cold 0.25 M sucrose and 0.02 M potassium chloride solutions into it.
- ☆ The samples are allowed to equilibrate for 5 min.
- ☆ Homogenize the samples for 40s at full speed with an Ultra Turrax tissue homogenizer.

- ☆ The homogenate is filtered through a pre-weighed muslin cloth through a filtration unit fitted with a funnel placed in a 50 ml test tube.
- ☆ Stir the homogenate with a glass rod to hasten filtration.
- ☆ Squeeze all the samples gently and uniformly in the muslin cloth to drain out the excess moisture present.
- ☆ The resulting fraction of muscle fragments collected on the screen is blotted with Whatman No. 1 filter paper.
- ☆ The weight of the sample with the screen is taken after 40 minutes of drying at 37°C in an incubator.
- ☆ MFI was calculated as a percentage of the weight of muscle fragments passed through (initial weight of muscle–weight of residue after drying) to that of the initial weight of the muscle sample.

6. Determination of pH of Meat (Trout *et al.*, 1992)

The normal physiological pH of muscular tissue ranges between 7.0-7.2. The ultimate pH of meat is reached to 5.4–5.6 in 18-24 hours post mortem. On storage, there is change in pH of meat and meat products due to metabolites of bacterial action. The deviation in pH is considered as indicator of the quality of meat sample. pH of meat is measured by different methods, either by digital pH meter or probe pH meter for online carcass pH determination or Nitrazine yellow indicator test or pH paper strip test. The pH of meat sample homogenate is measured either by electronic digital pH meter or by pH paper strips.

Reagents and Materials

1. Standard pH buffer solutions
2. Blender/pestle and mortar
3. pH meter
4. Beakers
5. Deionized water

Procedure

- ☆ Calibrate the pH meter using standard pH buffer solution (pH 4.0 and pH 9.0)

- ☆ The pH of meat sample is determined by blending 10 g of the sample with 50 ml distilled water for 1 min using pestle and mortar.
- ☆ The pH of the suspension is recorded by dipping combined glass electrode of a digital pH meter.

7. Water Holding Capacity

Method–I: Determination of Water Holding Capacity by Centrifugation Method of Wardlaw *et al.,* 1973

Water holding capacity (WHC) is defined as the ability of meat to retain its water during application of external force such as cutting, grinding, pressing, heating etc. Water liberated by such methods may be termed as loose water and the water retained by the tissue is called bound water. WHC can be expressed as amount of loose water related to the total muscle content in the muscle or the amount of bound water related to the muscle or muscle proteins. Water holding capacity of muscle tissue has a direct effect on shrinkage of meat during storage.

WHC is especially critical in meat products preparations that are formulated by combination of heating, grinding etc.

Reagents and Materials

1. 0.6 M NaCl solution
2. Minced Meat sample
3. Mechanical shaker
4. Polycarbonate Centrifuge tubes
5. Measuring cylinders
6. Glass rod
7. Refrigerated centrifuge

Procedure

- ☆ Take 100ml polycarbonate centrifuge tube
- ☆ Put 20 g of finely minced meat sample (W)
- ☆ Add 30ml of 0.6 M NaCl solution to it (V1)
- ☆ Mix with glass rod and stir for 2 min.

- ☆ Hold for 15 min at 4°C in a refrigerator in order to allow the effect of sodium chloride and phosphate to reach equilibrium.
- ☆ Shake the slurry for 1 min on a mechanical shaker
- ☆ Centrifuge the slurry at 5000 rpm for 15 min in a refrigerated centrifuge.
- ☆ Decant the supernatant fluid in a measuring cylinder and note the volume (V2)
- ☆ Calculate the difference between added NaCl solution and supernatant fluid as a percent of the weight of the meat sample.

Observations

Weight of meat sample = W (g)

Volume of 0.6N NaCl used = V1 (ml)

Volume of supernatant = V2 (ml)

Volume of 0.6N NaCl retained by sample = (V1 – V2) = V (ml)

Calculations

$$\text{WHC (per cent)} = V/W \times 100$$

Method–II: Determination of Water Holding Capacity (WHC) in Meat (Ham, 1960)

Reagents and Materials

1. 0.6 m NaCl
2. Refrigerated centrifuge
3. Graduated cylinder
4. Beakers

Procedure

- ☆ Weigh 15 g of ground meat, add 22.5 ml of 0.6 M NaCl, stir for 1 minute and let it stand for 15 minutes at 4°C,
- ☆ Stir again the meat slurry for 1 minute and immediately centrifuge at 10,000 rpm for 15 minutes in refrigerated centrifuge at 4°C.
- ☆ Measure the volume of supernatant.

- ☆ Record the amount of added solution retained by meat as WHC in ml/100 g of meat.

Method–III: Determination of Water Holding Capacity (WHC) in Meat

Water holding capacity (WHC) No. 1-Filter paper method developed by Grau and Hamm (1953, 1957) and modified by the National Animal Industry Experiment Station in Japan.

Materials

1. Filter papers
2. Glass slides
3. Pressing weight

Procedure

- ☆ Intact muscles (10mm × 0mm × 5mm) were pressed between two filter papers using a 35kg wt/cm^2 for one min.
- ☆ The amount of water released from the sample was measured indirectly by measuring the area of filter paper wetted relative to the area of pressed sample.

Calculations

The WHC of the meat was calculated as follows:

(Area of water released from meat (cm^2) – Area of meat sample (cm^2) × 9.47 ×100

WHC= 100 – Weight of meat sample (mg) × Moisture content of meat per cent

Method–IV: Determination of Water Holding Capacity (WHC) in Meat, Press Method

The filter paper press method is modified by Gnanasambandam and Zayas (1992) as given below:

Materials

1. Digital balance
2. Meat sample
3. Glass plates or Polyethylene sheets

4. Weights (18.5 kg)
5. Whatman filter paper No.1

Procedure

- ✰ Weigh two Whatman filter paper No. 1 (A)
- ✰ Weigh exactly 0.5g of meat sample. (B).
- ✰ The weighed meat sample is placed between the centre of two weighed filter papers.
- ✰ Polyethethylene sheets or glass plates are kept below and above the filter papers with sample.
- ✰ Pressure is applied by keeping weights of 18.5 kg on it for 5 min.
- ✰ Remove the weights and weigh the meat flakes (C)
- ✰ Filter papers are dried at room temperature and weighed again (D).

Observations

Weight of empty filter papers: A =

Weight of meat sample: B =

Weight of meat flakes: C =

Weight of dried filter paper: D =

Weight of meat residues adhered to the filter paper: E = D – A

Actual weight of meat sample: F = E + C

Calculations

$$\text{Water Holding Capacity (per cent loose water)} = \frac{B - F}{B} \times 100$$

Method–V: Water Holding Capacity: Hamm Press method (Hamm, 1960)

Reagents and Materials

1. Whatman No. 2 filter paper
2. Plexi glass plates
3. Saturated Potassium chloride solution
4. Planimeter

Procedure

- ✰ A 0.3g sample of meat batter was placed on a filter paper (Whatman No. 2, stored over saturated KCl; placed between two plexi glass plates and pressed for 20 min by 1 kg.
- ✰ Areas of moisture spread and meat batter were measured with a compensating polar planimeter.

WHC was determined using the equation

$$\text{WHC} = 1 - \frac{\text{Area of moisture}}{\text{Area of batter}}$$

Method–VI: Water Holding Capacity/Expressible Juice

A press technique improved by Wierbicki and Deatherage (1958) to measure free water in uncooked meat was used to determine percentage of expressible juice in the products.

Procedure

- ✰ Samples (500 mg) were placed on filter paper (Whatman No.1), which was positioned between two plexiglass sheets and pressed for 1 min at 5000 psi by a hydraulic press.
- ✰ The total wet juice area and meat area (6.45 cm^2) were measured with a planimeter.
- ✰ The percentage of expressible juice was calculated by the following equation:

$$\text{Per cent expressible juice} = \frac{\text{(Total juice and meat area – Meat area) (61.10) (mg/6.45 cm}^2\text{)}}{\text{Total moisture (mg) of meat sample}}$$

The coefficient of 61.10 was determined by Wierbicki and Deatherage (1958) and equals mg water absorbed per 6.45 cm^2 of Whatman filter paper No. 1 filter paper. Three samples from each treatment were tested.

8. Determination of Extract Release Volume (ERV) of Meat

ERV technique first described by Jay (1964) has been shown to be of value in determining spoilage in meats as well as in predicting refrigerated storage shelf life. ERV refers to volume of aqueous extract released by meat/fish homogenate when it is passed through a Whatman filter paper no. 1 for a given period of time. As meat undergoes microbial spoilage, there is a complete hydrolysis of proteins, which significantly decreases the ERV. Therefore, good quality meat has higher ERV value than poor quality meat. By this method beef of good organoleptically and microbial quality releases more volumes of extract as compared to poor quality beef. Ground meat failing to produce ERV of 25-30 ml should be regarded as spoiled and this corresponds to bacterial load of approximately 10^8cfu/g.

ERV during storage is a useful technique for assessing freshness of beef (Pearson 1968,)

Principle

ERV refers to volume of aqueous extract released by meat/fish homogenate when it is passed through a Whatman filter paper No. 1 for a given period of time. As meat undergoes microbial spoilage, there is a complete hydrolysis of proteins, which significantly decreases the ERV. Therefore, good quality meat has higher ERV value than poor quality meat.

Reagents and Materials

1. Stock Solution of phosphate buffer (0.2 M)
2. Working solution of phosphate buffer (0.05 M)
3. Minced Meat sample,
4. Tissue homogenizer/Pestle and Mortar.
5. Whatman Filter paper No. 1
6. Measuring cylinder
7. Funnel
8. Time clock
9. Glass beaker
10. Double distilled water/Normal saline solution.

☆ *Stock Solution of phosphate buffer (0.2 M)*: Mix solutions of 0.2 M $Na_2\,HPO_4$ (Mol. wt. 178) and 0.2 M $Na\,H_2PO_4$ (Mol wt. 156.1) in ratio of 1:8 and store in refrigerator

- *Working solution of phosphate buffer (0.05 M)*: Dilute 250 ml of stock solution to 1000 ml with distilled water to get 0.05 M phosphate buffer of pH 5.8.

Procedure

- Take 15g of minced meat and blend with 60ml of 0.05M PBS (pH 5.8) for two minutes in a tissue homogenizer.
- Filter the homogenate through Whatman No. 1 filter paper.
- Filtration is carried out for 15 min and filtrate is collected in a measuring cylinder.
- Volume of filtrate is taken as a measure of ERV.

Precautions

- Properly homogenize the sample.
- Note the time correctly as results are interpreted on the basis of time

Interpretation

- The good quality meat exudes 25-40 ml of ERV in 15 min however; poor quality/spoiled meat exudates less than 25 ml.

9. Determination of Cooking Release Volume (CRV) of Meat

Materials

1. Hot water bath
2. Polyethylene bags

Procedure

- Cook the meat in a constant temperature water bath at 80°C for 20 minutes by keeping them in a polyethylene bag.
- Calculate the weight loss during cooking by subtracting weight of meat after cooking from that before cooking and record it as C.R.V.

Calculations

CRV = Weight before cooking – Weight after cooking

10. Cooking Loss, Cooking Yield and Frying Loss

(*a*) Cooking Loss

Apparatus

- ✰ Autoclave

Procedure

- ✰ Take about ten 0.5 inches cubes of meat sample.
- ✰ Weigh the cubes
- ✰ Add 1 per cent salt
- ✰ Cook under 15 lbs pressure for 15 min.
- ✰ Cool the cooked meat to room temperature.
- ✰ Discard the fluid portion of the meat.
- ✰ Dry the cooked meat pieces by help of blotting paper without pressing.
- ✰ Weigh the cooked and dried cubes.

Calculations

$$\text{Cooking loss (per cent)} = \frac{W_1 - W_2 \times 100}{W_1}$$

where,

W_1= Weight of meat cube taken (g),

W_2= Weight of cooked and dried cube

Observations

Sample No.	*Weight of Meat Cube W_1*	*Weight of Cooked and Dried Cube W_2*	*Cooking Loss (per cent)*
1			
2			
3			

(*b*) Cooking Yield

The weight of meat blocks before and after cooking was recorded. Cooking yield was expressed in percentage.

(*c*) Frying Loss

The buffalo meat nuggets were fried with refined mustard oil following shallow pan-frying method till they get attractive golden brown color and served to the taste panel members for organoleptic evaluation. The weight of nuggets before and after frying was recorded. Frying loss was expressed in per cent.

11. Determination of Titrable Acidity/Total Acidity (Shelf and Jay, 1970)

Total acidity or titrable acidity expressed as lactic or acetic acid may be used as an indirect measure of bacterial growth in brines and liquid foods. Measurement of acid present may be accomplished by titration of a suitable quantity of sample with 0.1 N NaOH to the phenolphthalein end point (pH 8.3). Total acidity is the amount of standardized 0.1N NaOH used to neutralize the acidity of a meat sample to the pH level of phenolphthalein used as indicator.

In determining the organic acids in foods, titrable acidity has been repeated to be of more value than pH alone, as the latter is a measure of hydrogen ion concentration and organic acids do not ionize completely. It has further been reported that in measuring the titrable acidity the amount of acid that is capable of reacting with a known amount of base is determined (Jay, 1987).

When first signs of organoleptic spoilage appeared; mean volume of titrable acid nearly doubled from 1.32 to 2.58 ml of 0.02 N HCl. Based on a log bacterial counts of 8.5 and ERV of 25, volumes of acid <2.0 ml indicated meats of acceptable quality.

Principle

Total acidity or titratable acidity expressed as lactic or acetic acid may be used as an indirect measure of bacterial growth. Measurements of acid present may be accomplished by titration of a suitable quantity of sample with 0.1 N NaOH to the phenolphthalein end point (pH 8.3).

Reagents and Materials

1. 1 per cent Phenolphthalein indicator
2. 0.1N NaOH
3. Waring blender/pestle and mortar
4. Volumetric flask 250 ml

5. Whatman's filter paper No. 2
6. Burette 50 ml
7. Conical flask 100ml

Procedure

- ✰ Ten gram of meat sample was blended in 200 ml of distilled water and made up to 250 ml in volumetric flask.
- ✰ Slurry was filtered through Whatman filter paper No.2.
- ✰ 25 ml of filtrate was collected to which 75 ml distilled water was added with three drops of 1 per cent phenolphthalein indicator solution and titrated against 0.1N NaOH.
- ✰ The amount of 0.1 N NaOH required was used to calculate total acidity.

Observations

Sl.No.	*Initial Burette Reading*	*Final Burette Reading*	*ml of NaOH Consumed*
1.			
2.			
3			

Calculations

$$\text{Titrable acidity (\% lactic acid)} = \frac{\text{ml of 0.1 N NaOH} \times 0.1 \times \text{meq wt. of lactic acid} \times 100}{\text{Weight of sample in g}}$$

ml of 0.1N NaOH consumed during the titration =

$$\text{meq. wt. of lactic acid} = \frac{90}{1000} = 0.09$$

12. Determination of *o*-Tyrosine

o-Tyrosine has been proposed as an indicator of food irradiation. When phenylalanine is irradiated with gamma rays it is oxidized to give o- and m- tyrosine isomers. The conversion yield is proportional to the absorbed dose and temperature during irradiation. As many foods contain a constant level of phenylalanine in proteins, the level of *o*-tyrosine may be a good indicator of food irradiation. Nevertheless, the low levels of naturally occurring *o*-tyrosine present in some foods should be taken into account. The application is especially common in chicken, fish and shrimps. Methods to determine *o*-tyrosine should be highly sensitive and selective such as GC-MS, HPLC with fluorescence detection, HPLC-LIF and HPLC-ED

13. Tyrosine Value

Most of the tyrosine in meat is produced due to hydrolysis of meat proteins. Hydrolytic changes in meat during storage can be caused either due to bacterial proteolysis or automatic changes by inherent tissue enzymes. Jay (1987) mentioned that determination of tyrosine complex as one of the method for detecting microbial spoilage in meats, poultry and seafoods. Tyrosine value increase with advanced spoilage but was not very sensitive to changes during early stage of spoilage.

The procedure prescribed by Strange *et al.* (1977) was used with slight modification

Reagents and Materials

1. 20 per cent Trichloroacetic acid solution
2. 0.5N sodium hydroxide
3. Folin and Ciocalteu's reagent
4. Spectrophotometer
5. Whatman No.42 filter paper
6. Distilled water

Procedure

- ✰ Trichloroacetic acid (TCA) extract was prepared by blending 20g minced meat sample with 50 ml of precooled 20 per cent TCA solution for 2 min.
- ✰ After homogenization, the contents were transferred quantitatively to a beaker by rinsing with 50 ml cold distilled water, mixed and filtered through a Whatman's filter paper No. 42.
- ✰ To estimate tyrosine value, 2.5ml TCA extract was mixed with equal amount of distilled water.

- ✰ The mixture was then blended with 10 ml of 0.5 N Sodium hydroxide.
- ✰ Then 3 ml of diluted Folin and Ciocalteu's reagent (1 volume concentrated F and C reagent + 2 volumes distilled water) added to it and again shaken.
- ✰ The mixture was allowed to stand in a dark place at room temperature for 15 min. for proper color development.
- ✰ The optical density was measured at 700 nm by a spectrophotomer.
- ✰ Tyrosine value was calculated as mg tyrosine per g of meat by referring to a standard graph, which was prepared as per the procedure described by Pearson (1968c).

14. Emulsifying Capacity

Method–I

Alternative procedure suggested by Swift *et al.* (1961) and based on the principle of Pierce and Kinsella (1978).

Reagents and Materials

1. Ultra turrex Homogenizers
2. Whatman filter paper No. 1
3. Funnels
4. Refrigerated Centrifuge
5. 1M NaCl
6. Soy Refined Oil

Procedure

- ✰ Two gm of meat was blended with 100 ml of chilled 1M NaCl in a Ultra Turrex homogenizer for 2 minute
- ✰ The slurry was then centrifuged for 5 minute and supernatant was filtered through Whatman No.1 filter paper.
- ✰ To 30 ml of this filtrate 10 ml of soy refined oil was added.
- ✰ Emulsion was prepared in ultra turrax homogenizer for 2 minute.
- ✰ The emulsion was kept in the refrigerator at 4°C for 15 minute and then centrifuged for 1 minute.

- ☆ Protein content of the protein matrix before and after forming emulsion was determined by the procedure of Lowry *et al.* (1951).
- ☆ Emulsifying capacity was expressed as ml of oil emulsified per mg of protein.

Method–II: Estimation of Emulsifying Capacity (EC)

The method used for determining the EC was similar to that of Swift *et al.* (1961).

Reagents and Materials

1. 1M sodium chloride solution
2. Weigh scale
3. Ultra turrex homogenizer
4. Centrifuge machine
5. Separatory funnel
6. Refined mustard oil
7. Plastic beaker

Procedure

- ☆ Take finely minced 25 g meat sample.
- ☆ Homogenize it with 100 ml of cold 1M NaCl (5°C) for 1.5 minutes using Ultra Turrax tissue homogenizer at high speed (>13500rpm).
- ☆ Take 12.5 g of the resulting slurry and add 37.5 ml cold 1M NaCI .
- ☆ Transfer it to 250 ml plastic beaker and homogenized again for 5 seconds at low speed (1000 rpm).
- ☆ To this add 50 ml of refined mustard oil and subjected to high speed cutting-mixing at approximately 13500 rpm, in Ultra Turrax tissue homogenizer and simultaneously add refined vegetable oil at a rate of about 0.8 ml per sec from a graduated separatory funnel
- ☆ An emulsion formed, persisted and finally collapsed.
- ☆ The transition is noticed by a gradual increase followed by a sudden decrease in viscosity.

- ✰ The addition of oil is immediately terminated on observation of the abrupt transition. The end point is also indicated by a change in sound in the pitch of homogenizing beaker.
- ✰ The volume of oil (50 ml and the additional oil drawn from the separatory funnel) just exceeded the EC of the meat sample is recorded and expressed as ml of oil emulsified per 2.5 g meat.

Method–III: Emulsion Capacity (EC)

EC, an indicator of the functional property of meat in an emulsion system, was determined by using a model system described by Ockerman (1976). The method utilized for end point determination was described by Webb *et al.* (1970).

Reagents and Materials

1. Sodium chloride
2. Potassium Hydrogen Phosphate
3. Blender
4. Ohm meter
5. Burette
6. Centrifuge
7. Kjeldahl assembly

Procedure

- ✰ NaCl and K_2HPO_4 at the indicated percentages were combined and dissolved in water.
- ✰ To measure EC, 25g of ground meat and 100 ml of cold (0-4°C) Salt-Phosphate solution (SP) were placed into a blender (Waring blender model 34B 199) jar and comminuted for 3 min at 13,000 rpm.
- ✰ Slurry 12.5 g and 37.5 ml of additional SP solution were transferred to another blender jar and homogenized 10 s at low speed (5000 rpm).
- ✰ Then 50 ml of corn oil was added to detect break point of emulsions, electrodes were connected to an Ohm meter (Huang Chang HC-3010 B2).
- ✰ Oil was added at @0.7ml/sec.

- ✰ The blender rate was 13000rpm during emulsification.
- ✰ The water-jacketed burette containing oil was maintained at the specified temperature.
- ✰ The emulsion breakpoint occurred when the Ohmmeter connected to a milli volt recorder (Karl Kolb RE 541) showed a sudden increase in resistance.
- ✰ At break point, oil addition was stopped.
- ✰ The total amount of oil emulsified included the first 50 ml added.
- ✰ EC was calculated as ml oil/g protein after determining the protein (Kjeldahl) content of the meat samples. (Ockerman, 1976)

15. Emulsion Stability (ES)

The method was based on the emulsion stability test reported by Townsend *et al.* (1968), Baliga and Madaiach (1970) and Parks and Carpenter (1987) with some modifications.

Materials

1. Low Density Polyethylene bags of150 gauge (size 11 × 10 cm)
2. Hot water bath
3. Weighing balance
4. Meat sample

Procedure

- ✰ About 25 g of raw emulsion samples were placed in low density polyethylene (LDPE) bags of 150 gauge (size 11 × 10 cm)
- ✰ Bags with samples were weighed and sealed.
- ✰ These bags were placed in a thermostatically controlled water bath and samples cooked at 80°C for 20 minutes.
- ✰ The bags were removed from water bath, cut open and the cook fluid (fat, water soluble solids) drained.
- ✰ The cooked samples were weighed.
- ✰ Loss in weight was calculated and expressed (in per cent) as an index of emulsion stability.

16. Thiobarbituric Acid (TBA)/Thiobarbituric Acid Reactive Substances (TBARS) Value

TBA value is defined as the increase in absorbance measured at 530 nm due to the reaction of the equivalent of 1 mg of sample per 1 ml volume with 2-thiobarbituric acid It is based on the principle that secondary oxidation products of oils and fats react with 2-thiobarbituric acid forming condensation products, the absorbance of which is measured at 530 nm, the wavelength of one of their absorption maxima. Alternatively, the condensation products may also be determined spectrofluorimetrically. Although indirectly, this value may also provide information on the presence of oxidized fatty acids.

Method–I: Determination of Thiobarbituric Acid Reactive Substances (TBARS), Tarladgis *et al.*, 1960

Principle

The sample is digested in acidic media and separated by using specific reflux distillation. Animal tissues which had been incubated aerobically produce a color with 2-thiobarbituric acid (TBA). This color is the result of a complex formed from oxidation products of unsaturated fatty compounds and 2-thiobarbituric acid. The primary products of lipid oxidation are hydroperoxides and these are readily decomposed to produce various thiobarbituric acid-reactive substances (TBARS) particularly carbonyl compounds. The 2-thiobarbituric acid reaction has been widely used as an objective measure of oxidative deterioration occurring in oils, fatty foods and correlates reasonably well with taste panel results. The compounds in distillate are estimated by spectrophotometric measurement at 538 nm after the reaction with 2-thiobarbituric acid. This technique is simple and reproducible with a detection limit of 0.2 N mol TBARS per 10 g of tissue sample and an overall deviation of less than 7 per cent.

Reagents and Materials

1. 2-thiobarbituric acid
2. Propyl gallate
3. Disodium EDTA
4. Anti-bumping granules

5. Standard 1, 1, 3, 3–Tetraethoxypropane (TEP)
6. Vertical distillation assembly*
7. A vertics 23 blender.
8. Hot water bath.
9. Spectrophotometer
10. 15 × 125 mm screw-cap test tubes with teflon-lined caps.

* Vertical distillation assembly, which keeps a constant reflux ration and gives reliable separation of TBARS (digest and separate TBARS).

TBA Reagent

2-thiobarbituric acid (1.44g) and 50 ml of distilled water were mixed in a 500 ml vol. flask with vigorous stirring (Magnetic stirrer). Glacial acetic acid was then added untill the flask was two-thirds full. The mixture was vigorously stirred for 10 minutes or until the 2-thiobarbiturtic acid was almost completely dissolved. The flask was then filled to the mark with glacial acetic acid.

TEP Standard Solution

1, 1, 3, 3–Tetraethoxypropane (TEP) (0.220 g) was accurately weighed into a 100 ml vol. flask and diluted to volume with distilled water to produce a 1×10^{-4} M. Working solution was then prepared by diluting 10 ml of the stock solution to 100 ml.

Preparation of Standard Curves

Aliquots of 0, 0.4, 0.8, 1.2, 1.6 and 2.0 ml of working TEP standard solution were accurately pipetted into screw cap test tubes and water was added to a total volume of 5 ml. TBA reagent (5 ml) was added and the tubes tightly capped. The final concentration of TEP in the above 10ml volumes corresponded to 0, 4.0, 8.0, 12.0, 16.0 and 20.0×10^{-7} mol per litre respectively. After thoroughly mixing, the test tubes were heated in a vigorously boiling water bath for 45 minute and cooled in tap water. Absorbance of the solutions was determined at 538 nm within 30 minutes of cooling, setting the blank (0.0 ml TEP) to zero. The plot of TEP concentration (μM) against absorbance at 538 nm was linear up to 2.0 μM TEP under the condition described here. The molar extinction coefficient of the color developed (absorbance/molarity) was 1.9×10^5 at 538 nm.

Procedure

TBARS Distillation

- ✰ Pre-weighed 10g meat samples are finely chopped in a food process or blender were placed in small containers and frozen immediately.
- ✰ Without thawing, a 10g portion of meat was transferred to the blender jar with 35 ml of distilled water and blended for 2 minute or until the sample was firmly divided.
- ✰ The sample homogenate was transferred to a tared 500 ml round bottom flask containing approximately 100 mg each of propylgallate and EDTA and distilled water was added so that the total weight of the sample and water was 105 g.
- ✰ The sample was flushed with nitrogen and 95 ml 4M HCl were added.
- ✰ The distillation was started immediately and 50 ml distillate was collected in a volume flask within 35 minute or less.
- ✰ The distillation rate was kept at about one to two drops per second.
- ✰ The still, between samples, was rinsed with methanol and then distilled water.
- ✰ TBARS distillates may be refrigerated overnight if necessary.

Spectrophotometric determination

- ✰ Samples (5 ml) of each TBARS distillate and 5 ml of TBA reagent were pipetted into screw-cap test tubes covered tightly and then treated as described in standard curve preparation.
- ✰ A blank of 5 ml of distilled water and 5 ml of TBA reagent was run simultaneously.
- ✰ Sample solutions with absorbance greater than 0.5 were diluted with distilled water or alternatively the analysis repeated using less TBARS distillate.

Calculations

TBARS value is expressed as μmol malonaldehyde per kg meat.

If a 5 ml aliquot of distillate obtained from 10g meat was used, then TBARS value may be calculated from the simplified formula:

$$C \times 10^7 = TBARS \qquad (1)$$

where, C (1.68×10^{-6}) represent equivalent concentration in mol per litre of TEP determined from the standard curve.

For aliquot other than 5 ml, the formula becomes :

$$\frac{5}{\text{Aliquot size}} \times C \times 10^7 = TBARS \qquad (2)$$

Method–II: Thiobarbituric Acid (TBA) Value

Method of American Oil Chemist's Society described by Koniecko (1979) and modified by Ke et al, (1984) was followed.

Reagents and Materials

1. Sulphanilamide
2. Distilled water
3. Kjeldahl flask
4. Paraffin wax
5. Distillation Assembly

Procedure

- ✰ 10g sample blended and 49 ml D.W. and 1 ml sulphanilamide reagent was added.
- ✰ The meat quantitatively transferred into a Kjeldahl flask.
- ✰ Another 48 ml D.W. was used for rinsing the blender and poured into the flask followed by addition of 2 ml of dil HCl (1 vol HCl + 2 vol DW).
- ✰ Few chips of paraffin wax added and flask heated for distillation .
- ✰ Rest procedure same as Tarladgis *et al*. (1960).

Method–III: TBA Value (Witte *et al.,* 1970)

Reagents and Materials

1. Trichloroacetic Acid (20 per cent)
2. Phosphoric Acid (2M)

3. Distilled water
4. Whatman Filter Paper No. 1
5. Test tubes
6. Ultra Turrex Homogenizer/Blender
7. Volumetric Flask
8. Spectrophotometer

Procedure

- ✰ 20 g sample homogenized with full speed for 1.5 minute in a chilled stainless steel waring blender cup with 50 ml extracting solution containing 20 per cent trichloroacetic acid in 2M phosphoric acid.
- ✰ The temperature of extraction solution was also maintained at 4°C.
- ✰ The resulting slurry was transferred quantitatively to a 100 ml vol. flask with 40 ml water. The volume was made up to 100 ml and thoroughly mixed by shaking.
- ✰ Half of the portion was filtered through Whatman No. 1 filter paper.
- ✰ 5 ml of filtrate was transferred to a test tube (15 × 200 mm) to which 5 ml of 2-thiobarbituric acid (0.005 M in distilled water) was added.
- ✰ The tube was stoppered and the solution mixed by inversion and kept in dark for 15 hours at normal temperature.
- ✰ The resulting color was measured at 530 nm in a spectrophotometer.

Calculations

The TBA value was calculated by multiplying the absorbance by the K-value of 5.2.

Method–IV: Estimation of Thiobarbituric Acid (TBA) Value by Extraction Method of Witte *et al.*, 1970 with Slight Modifications

Thiobarbituric Acid (TBA) value/Thiobarbituric Acid Reactive Substances (TBARS) number is used as an indicator of oxidative rancidity in meat foods. It is highly correlated with oxidized and warmed-over-flavour of meat and meat products.

Principle

Malonaldehyde reacts with Thiobarbituric acid and form a conjugate, which on high temperature gives rise to red color which is estimated in a spectrophotometer by recording OD 532nm and the TBA value is expressed as mg malonaldehyde per kg of meat.

TBA assay measures the amount of a secondary metabolite produced mainly from oxidation of polyunsaturated fatty acids in meat sample.

Reagents and Materials

1. 1,1,3,3 tetra ethoxypropane (1 mg/ml) stock solution
2. 1,1,3,3 tetra ethoxypropane (3 µg/ml) working solution
3. TBA reagent (1 mg/ml)
4. 20 per cent Trichloroacetic acid
5. Meat sample
6. Meat blender/homogenizer/pestle and mortar
7. Hot Water bath
8. Distilled water
9. Graduated cylinder/volumetric flask
10. Test tubes with stand
11. Spectrophotometer

Procedure

- ☆ Take 10 g of meat sample and blend it finely with 50 ml of 20 per cent TCA in a waring blender/homogenizer for 2 minutes. The pestle and mortar can also be used for homogenization of sample.
- ☆ The resulting slurry was allowed to stand for 10 min.
- ☆ Filter the extract through Whatman filter paper No. 42.
- ☆ In a test tube take 3 ml of this extract and mix with equal volume of 0.1 per cent (w/v) TBA reagent.

- Prepare blank sample by mixing 3 ml 20 per cent TCA with equal volume of 0.1 per cent TBA reagent
- The content of each test tube is thoroughly mixed and boiled for 35 minutes in a boiling water bath.
- Cool the test tubes.
- Measure the absorbance at 532 nm by a spectrophotometer.
- TBA value is calculated by comparing the absorbance of test sample with a standard graph prepared by using known concentrations of malonaldehyde.
- For preparing standard graph, dissolve 0.3055 gm of 1,1,3,3 Tetraethoxy Propane (TEP) in 100 ml of 95 per cent absolute alcohol to obtain a concentration of 1mg malonaldehyde/ml. This is stock solution.
- To prepare working standard solution of TEP, 0.3 ml of stock solution is diluted to a volume of 100 ml by distilled water. The diluted solution contains 3 μg/ml of malonaldehyde.
- From this solution a standard graph is prepared by using different concentration of malonaldehyde and TBA value was estimated.

Precautions

- Properly blend the meat sample with TCA.
- Run blank along with meat sample with similar treatment.

Method–V: Determination of TBA Value by Distillation Method

Principle

Same as TBA estimation by extraction method described earlier.

Reagents and Materials

1. Thiobarbituric acid reagent (1 mg/ml).
2. 6N HCl
3. Meat sample
4. Waring Blender/homogenizer/pestle and mortar
5. Volumetric flask 100ml

6. Water bath
7. Glass stopper tubes 50 ml cap.
8. Spectrophotometer
9. Paraffin wax
10. Glass beads
11. Whatman's filter paper No.1
12. Kjeldahl distillation apparatus
13. Kjeldahl distillation flask 500 ml
14. TEP stock solution
15. TEP Working solution

- *1,1, 3, 3, tetra ethoxy propane (1mg/ml) stock solution*: Prepared by dissolving 0.3055 g of TEP in 100 ml of 95 per cent alcohol.
- *Working standards of TEP (ranging from 0.6μg to 1.8μg/ml)*: Prepared by taking 0.3 ml of stock solution of TEP and making the volume to 100 ml by distilled water, which contained 3μg/ml.

Procedure

- Working standards ranging from 0.6 μg to 1.8 μg were prepared as per TBA estimation by extraction method described earlier.
- Test tube No. 4 was taken as blank sample.
- The unknown meat samples were prepared for distillation and got ready for recording Optical Density as described below.
- 10 g of meat was blended with 50 ml distilled water in a waring blender for 2 minutes. The mixture was transferred quantitatively into a Kjeldahl flask of 500 ml capacity by washing with additional 45 ml distilled water. 5 ml of 6N HCl solution was added to it to bring the pH to 1.6.
- Small amount of paraffin wax and few glass beads were put in the Kjeldahl flask in order to prevent bumping.
- The apparatus was assembled and the flask was heated at the lowest heat obtainable on the Kjeldahl distillation apparatus and 50 ml of distillate were collected.

☆ Distillates were thoroughly mixed and 5 ml of distillate were filtered to avoid precipitate if any, into a 50 ml glass stoppered tube and to this 5 ml of TBA reagent was added. Tubes were stoppered, contents mixed and immersed in a boiling water bath for 35 minutes.

☆ A distilled water blank was also prepared and heated like the sample.

☆ After heating, the tubes were cooled in tap water for 10 minutes and portions of these were transferred to a cuvette.

☆ Optical Density of the sample was read against blank at 538 nm.

☆ Readings were noted and with the help of "Standard TEP curve" mgs of malonaldehyde per kg of meat was calculated.

Calculations

TBA value = O.D. at 538nm × 7.8 (factor)

= mg malonaldehyde/kg

Method–VI: Thiobarbituric Acid (TBA) Number

Thiobarbituric acid numbers were determined on cooked white and dark meat samples on days 0,5,10 and 15 as described by Salih *et al.* (1987) with some modifications.

Reagents and Materials

1. Perchloric Acid
2. Beta Hydroxy Toluene
3. Whatman No. 1 filter paper
4. 20mM TBA reagent
5. Distilled water
6. Spectrophotometer

Procedure

☆ Meat samples 2 g were weighted into 50 ml test tubes and 18 ml of 3.86 per cent perchloric acid was added.

☆ The samples were homogenized for 15 sec at high speed. BHT was added (125 mg/g fat) to each sample during homogenization to control lipid autooxidation.

- ☆ The homogenate was filtered through Whatman No. 1 filter.
- ☆ Filtrate (2ml) was mixed with 2 ml of 20 mM TBA in distilled water and incubated at room temperature (23°C) in the dark for 17 h.
- ☆ Absorbance was determined at 531 nm against blank containing 2 ml distilled water and 2 ml of 20 mM TBA solution.
- ☆ The TBA number was expressed as mg malonaldehyde/ kg meat.

Method–VII: Modified Distillation Method

Determination of Malonaldehyde (TBA Values) as an Index of Rancidity in Meat

Reagents and Materials

1. *TBA reagent*: 0.02 M 2-thiobarbituric acid in 95 per cent glacial acetic acid. Bring into solution by warming slightly in a boiling water bath.
2. Malonaldehyde tetra ethyl acetal standard–1, 1, 3, 3, tetraethoxypropane (TEP):
   (*a*) Stock solution A-Dissolve 0.3055 g TEP in 100 ml 95 per cent alcohol. This solution contains 1.000 mg/ml as malonaldehyde and can be kept several days in refrigerator and diluted when necessary.

   (*b*) Stock solution B-Transfer 10 ml stock solution A to 250 ml volumetric flask, dilute the volume with distilled water and mix. This solution contains about 40 µg/ml as malonaldehyde.
3. Acetic acid 90 per cent.
4. HCl-1.5 N
5. Antifoam A-Dow corning
6. Spectrophotometer
7. Distillation assembly-Kjeldahl distillation rack with 500 ml flasks.

Procedure

Preparation of Standard Curve (Holland, 1971)

- Pipette out 1.0, 2.0, 3.0, 4.0 and 5.0 ml stock solution B into each of five 200 ml volumetric flask and dilute the volume with distilled water (working standards range from about 0.2 to 1.0 µg/ml).
- Transfer 5 ml from each of the standard solutions to 50 ml glass stoppered tubes.
- Add 5 ml of TBA reagent, mix thoroughly and immerse tubes in boiling water bath until maximum color is developed (about 35 minutes at 95°C).
- Cool tubes to room temperature and determine absorbance of each standard solution and blank by scanning from 600-400 nm (maximum at 532 nm).
- Plot absorbance vs. µg malonaldehyde/ml final solution. Final solutions range from 0.1 to 0.5 µg/ml.

Distillation of Meat Sample (Tarladgis, 1960)

- Blend 10 g of meat with 50 ml of distilled water in warring blender for 2 minutes.
- Transfer the mixture quantitatively into a Kjeldahl flask by washing with an additional 47.5 ml of distilled water.
- Add 2.5 ml of HCl solution to bring the pH to 1.5. Place a small amount of antifoam A in the flask and a few saddle stones to prevent bumping.
- Assemble apparatus and heat flasks at the lowest heat obtainable on the Kjeldahl distillation apparatus with the electric heating element.
- Approximately 10 minutes from the moment boiling begins are required to collect 50 ml of distillate.
- Mix the distillate, pipette 5 ml into a 50 ml glass stoppered tube and add 5 ml of TBA reagent.
- Stopper the tube, mix the content and immerse in a boiling water bath for 35 minutes.
- A distilled water TBA reagent blank should be prepared and treated like the sample.

☆ After heating cool in tap water for 10 minutes and take the reading at 538 nm.

☆ Multiply the reading by factor 7.8 to convert to mg of malonaldehyde/1000 g of meat.

Method–VIII: Accelerated TBARS Value

It is a measure of potential protection against lipid oxidation by the addition of hurdles to the meat products. In this method, lipid oxidation was accelerated by temperature (50°C) and the addition of $FeSO_4$. The procedure described by Juncher *et al.* (2000) was followed.

Reagents and Materials

1. Orthophosphoric acid
2. 5mM Phosphate Buffer
3. 15 per cent Trichloro acetic acid
4. Propyl gallate
5. Ferrous sulphate
6. Sulphanilamide
7. Standard thiobarbituric acid
8. Hot water bath
9. Spectrophotometer
10. Ultra turrex homogenizer
11. Shaker
12. Whatman's filter paper

Procedure

☆ A 15 per cent meat emulsion was prepared by blending 7.5 g product in 42.5 ml 5 mM phosphate buffer (T=50°C, pH=5.80, pH adjusted with orthophosphoric acid) using an Ultra Turrax tissue homogenizer (Model T25, Janke and Kenkel, 1 KA Labor Technik, Germany) at 13, 500 rpm for 1 min.

☆ Take 1.5 ml emulsion out of this and transfer it to a 10 ml test tube.

☆ Add 0.25 ml 3.74 m M $FeSO_4$ solution.

☆ The test tubes were placed in thermostat controlled water bath at 50°C and continuously shaken. TBARS was measured by addition of 6 ml 15 per cent trichloroacetic acid solution containing 0.1 per cent propyl gallate and 0.015 per cent sulphanilamide after 0 and 4 hr.

- ☆ The samples were vigorously shaken on Cyclo-mixer/ shaker for 20 s and filtered through a Whatman filter paper No. 42.
- ☆ The extract (3 ml) was mixed with 0.026 M thiobarbituric acid (2 ml) and heated in a 100°C water bath for 30 min followed by cooling in ice water for 5 min.
- ☆ The absorbance was measured at 532 and 600 nm on a spectrophotometer.
- ☆ The color developed after 4 h was also measured as described above.
- ☆ Blank was prepared as described above but using 1.5 ml 5 mM phosphate buffer instead of 1.5 ml emulsion.
- ☆ The difference in absorbance A_{532} nm–A_{600} nm (where A_{600} nm was used to correct for sample turbidity) was used as the measurement for the degree of lipid oxidation.
- ☆ All measurements made in triplicate and take average value.

17. Estimation of Free Radical Scavenging Activity

Reactive oxygen species including free radicals such as superoxide anion radical ($O_2^{-\cdot}$), hydroxyl radical ($OH^{\cdot}$) and singlet oxygen are generated as oxidation product of biological reactions or exogenous factors. These reactive oxygen species induce oxidative damage to biomolecules like lipids, nucleic acid, proteins, amines, deoxyribonucleic acid and carbohydrates and their oxidation products which cause aging, cancer, heart disease, stroke, arteriosclerosis, diabetes etc. Scavenging of free radicals by antioxidants help in calculating its antioxidant potential. It can be measured by various methods

17.1 ABTS Method

Free radical scavenging activity was determined by ABTS method (Pellagrani *et al.*, 1999; Hern´andez-Ledesma *et al*, 2005). In this assay, radical mono cation 2, 2′azinobis (3-ethylbenzothiazoline-6-sulfonic acid) ($ABTS^{+}$) are generated by potassium persulfate oxidation of ABTS and are reduced in the presence of hydrogen donating antioxidant, thus lowering absorbance at 734 nm.

Reagents and Materials

- *Potassium persulphate solution (140 mM)*: 1.892 gm of potassium persulphate was weighed and dissolved in double distilled water and made the volume to 50 ml.
- *ABTS [2, 2'-azinobis (3 ethyl benzothiazoline)-6-sulfonic acid] stock solution*: 19.2 mg of ABTS (Sigma-Aldrich) was weighed and dissolved in 5 ml of double distilled water, added 88 ml of 140 mM potassium persulphate solution and the mixture was kept in an Amber color bottle in dark for 12-16 hours.
- *Phosphate buffered saline (PBS)*: Dissolved 8 g of NaCl, 0.2 g of KCl, 1.44 g of Na_2HPO_4 and 0.24 g of KH_2PO_4 in 800 ml distilled water, adjusted pH to 7.4 with 1 N HCl and made the volume to 1 litre with distilled water.
- *ABTS working solution*: 1 ml of ABTS stock solution was diluted with phosphate buffer saline (approx 1:70) till it give an absorbance of 0.70±0.02, before that absorption spectra of ABTS was analyzed and maxima was at 734 nm.
- *Trolox solution*: 12.5 mg of Trolox [6-hydroxy. 2, 5, 7, 8–tetramethyl chroman-2-carbocyclic acid] (Sigma-Aldrich) was weighed and dissolved in 10 ml of ethanol. The resulting solution was 5 mM trolox solution. It was diluted with distilled water to 250mM concentration.
- UV-Visible spectrophotometer
- Microcuvette
- Micro pipette
- Regular glassware

Procedure

Preparation of Standard Curve

- 1 ml of ABTS working solution was taken in microcuvette and initial absorbance against buffer blank was recorded at 734 nm using double beam spectrophotometer.

- ☆ The appropriate volumes of Trolox (250 μM) solution (0,5,7.5,10,12.5,15,17.5, 20μl) corresponding to concentration of 5-20 μM were added to microcuvette using micropipette.
- ☆ The contents were mixed for 30 sec. and change in absorbance at 734 nm was recorded over 10 min.
- ☆ The standard curve was prepared by plotting concentration (μM) of Trolox (X-axis) v/s per cent inhibition (Y-axis).

Trolox Equivalent Antioxidant Capacity (TEAC)

- ☆ 1 ml of ABTS working solution made with PBS (pH 7.4) was taken into microcuvette (1 ml capacity) and absorbance adjusted to 0.70±0.02 against the buffer.
- ☆ 10μl of sample was added to ABTS working solution as well as in the blank.
- ☆ The contents were mixed for 5 seconds and change in absorbance at 734 nm was recorded over 10 min using double beam spectrophotometer.

Calculation

Based on the per cent inhibition of absorbance of sample, trolox equivalent was determined from standard curve, using following equation:

$$y = 3.86\,x + 2$$

where,

y is the per cent inhibition =

$$[(A\ 734\ nm_{control} - A\ 734\ nm_{sample}) / A\ 734\ nm_{control}] X\ 100$$

x is the mM concentration of trolox

The results were expressed as trolox equivalent antioxidant capacity (TEAC) values *i.e.* μmol of Trolox equivalence/mg of the protein.

17.2 Assay for DPPH Radical Scavenging Activity

The method of Shimada *et al.* (1992) was used to assess the DPPH radical scavenging activity of casein and its fractions.

Reagents and Materials

- *DPPH methanolic solution (1 mM)*: 4 mg of 2, 2-Diphenyl-1-picrylhydrazyl (Sigma) was dissolved in 10 ml of HPLC grade methanol (SRL).
- *Trolox Solution*: 12.5 mg of Trolox [6-hydroxy. 2, 5, 7, 8–tetramethyl chroman-2-carbocyclic acid] (Sigma-Aldrich) was weighed and dissolved in 10 ml of methanol. The resulting solution was 5 mM trolox solution. It was diluted to 250mM concentration.
- UV-Visible spectrophotometer
- Microcuvette
- Micro pipette
- Regular glassware

Procedure

- Five ml of the sample solution/trolox solution was mixed with 1 ml of freshly prepared DPPH methanolic solution (1 mM).
- The resulting solution was then mixed vigorously and incubated for 30 min at room temperature.
- The absorbance of the resulting solution was measured at 517 nm using double beam spectrophotometer.
- Under similar conditions, blank value was determined by using phosphate buffer (0.1 M, pH 7.0) and measured the reduction in absorbance.

Calculation

The results were expressed as

$$\%\text{DPH scavenging activity} = 1 - \frac{\text{Test sample absorbance}}{\text{Blank sample absorbance}} \times 100$$

And as (IC_{50}) value–the concentration of the sample required to scavenge 50 per cent of the free radicals.

17.3 Assay of Superoxide Anion Scavenging Activity (SOSA)

The method described by Liu *et al.* (1997) was used for the measurement of SOSA of casein and its fractions.

Reagents and Materials

- Phosphate buffer (0.1 M, pH 7.4): 19.0 ml of potassium dihydrogen phosphate (0.2M) was taken and 81.0 ml of (0.2 M) dipotassium hydrogen phosphate was added to it and volume was made to 200 ml using double distilled water.
- β-Nicotinamide Adenine Dinucleotide (NADH) solution (0.468 mM) : Dissolved 3.57 mg of NADH, Reduced (Amresco) in 10 ml of 0.1 M phosphate buffer.
- Nitroblue Tetrazolium solution (NBT) (1mM) :Dissolved 1.27 mg of NBT (Himedia) in 10 ml of 0.1 M posphate buffer, pH 7.4.
- Phenazine Methosulphate solution (PMS) (0.5 mM) : Dissolved 1.8 mg of PMS (Himedia) in 100 ml of 0.1 M phosphate buffer, pH 7.4.
- UV-Visible spectrophotometer
- Microcuvette
- Micro pipette
- Regular glassware

Procedure

- 1 ml of NBT solution, 1 ml NADH solution and 0.1 ml of sample solution were mixed.
- The reaction started by adding 100 µl of PMS solution to the mixture.
- The reaction mixture was incubated at 25°C for 5 min, and the absorbance at 560 nm was measured against blank samples.
- Decreased absorbance of the reaction mixture indicated increased superoxide anion scavenging activity.

The percentage inhibition of superoxide anion generation was calculated using the following formula:

$$\% \text{ Inhibition} = \frac{A_0 - A_1}{A_0} \times 100$$

where,

A_0 was the absorbance of the blank and A_1 was the absorbance of the samples.

18. Fat Quality Parameters

A. Saponification Value

Saponification value is the amount of alkali necessary to saponify a definite quantity of the lipid sample. It is expressed as the number of milligrams of potassium hydroxide (KOH) required to saponify 1 g of the sample. The saponification value gives information on the average chain length of the fat constituent fatty acids. It increases when the chain length decreases and conversely. For example, the saponification value of sunflower oil rich in C 18 fatty acids is about 190, while that of coconut oil rich in lauric acid (C12) is about 260.

B. Acid Value

The acid value is the number of milligrams of KOH necessary to neutralize the free acids in 1 g of sample. With samples that contain virtually no free acids other than fatty acids, the acid value gives information of the free fatty acids present in the sample.

C. Iodine Value

The iodine value is a measure of the unsaturated fats and oils and is expressed in terms of the number of centigrams of iodine absorbed per gram of sample (per cent iodine absorbed). Iodine value gives information on the average degree of unsaturation of the constituent fatty acids. It increases with unsaturation. For example, the iodine value of sunflower oi1 rich in linoleic acid (18:2 n-6) is about 125 while that of coconut oil rich in saturated fatty acids is only about 8.

D. Polyunsaturated Acids

Polyunsaturated acids may be calculated by the ultraviolet absorption in a purified solvent. If acids are conjugated, the absorption is measured directly. If acids are nonconjugated, they may be partially conjugated by heating in a potassium hydroxide-glycol solution and then the absorption of the conjugated constituents is measured in an ultraviolet spectrophotometer. The percentages of conjugated diene, triene, tetraene, and pentaene and of linoleic, linolenic, arachidonic, and pentaenoic acids are calculated from these measurements.

E. Hydroxyl Value

The hydroxyl value is defined as the number of milligrams of potassium hydroxide equivalent to the hydroxyl content of 1 g of sample. The hydroxyl value gives information on the presence or absence of hydroxylated fatty acids in fats. For example, castor, which is rich in ricinoleic acid, has a hydroxyl value of about 150.

F. Oxirane Oxygen Value

The oxirane oxygen value determines the presence of epoxy groups by direct titration of the sample with hydrogen bromide in acetic acid. This value provides information on the presence of epoxidized fatty materials and epoxy compounds in general.

G. Peroxide Value

The peroxide value determines all substances, in tends of milliequivalents of peroxide per kilogram of sample, that oxidize potassium iodide (KI) under the conditions of the test (132-135). The substances are generally assumed to be peroxides or other similar products of fat oxidation. Therefore, this value gives information on the presence of oxidized fatty acids.

H. *p*-Anisidine Value

This method determines the amount of aldehydes (principally 2-alkenals and 2,4-dialkenals) in animal and vegetable fats and oils, by reaction of these compounds with p-anisidine. Although it does not measure fatty acids, the presence of aldehydes may also provide information on the presence of oxidized fatty acids.

19. Determination of Refractive Index of Fat

Objective

- ☆ To estimate the refractive index of neats foot oil using Abbe's refractrometer.

Principle

Refraction of light rays differs for different media. This property can be used to determine the purity of the material. Abbe's refractrometer is the apparatus giving direct reading of refractive index. The R.I. of neats foot oil was 1.4516

Reagents and Materials

1. Distilled water
2. Refractometer
3. Prism Box
4. Neat's foot oil

Procedure

- ☆ For examination of liquid open the prism box and place a few drops of distilled water. Reading was taken and compared with the standard table of refractive index at various temperatures.
- ☆ Instrument correction factor negative or positive was used.
- ☆ Then placed a few drops of neats foot oil on the ground surface of lower purisms, closed and fastened the prism box.
- ☆ Forced the crosswire of eye piece and adjusted the mirror until the illumination is good and gently turn the prism box by means of a milled head screw forward and backward until the color band appear in the field of view.
- ☆ By turning the milled ring of the base of the telescope, a position will be obtained in the field of view, which is partly light and partly dark with the intersection of the crosswire.
- ☆ Read the refractive index on the divided scale by means of a magnifier. The scale is divided to third decimal part. The fourth decimal being easily obtained by estimation.

Standard Table for Refractive Index of Water

Temperature (°C)	*Refractive Index*	*Temperature (°C)*	*Refractive Index*
14	1.33348	32	1.33164
15	1.33341	34	1.33136
16	1.33333	36	1.33107
20	1.33299	38	1.33079
22	1.33281	40	1.33051
24	1.33262	42	1.33023
26	1.33241	44	1.32992
28	1.33219	46	1.32959
30	1.33192	48	1.32927

Observations

Refractive Index of water at 20°C =

Refractive Index of water as per standard =

Error =

Calculations

Refractive Index of neats foot oil =

20. Determination of Saponification Number

Method–I

Principle

Saponification number is the number of mg KOH required to saponify per 1g fat using equal weights. A fat of low molecular weight will require more KOH to saponify it than a fat of high molecular weight.

Reagents and Materials

1. Alcoholic Potassium hydroxide solution
2. Hydrochloric Acid (HCl)
3. Phenolphthalein indicator
4. ErlenMeyer flask
5. Air condenserAlcoholic Potassium hydroxide solution:

Reflux 1-2 litre of alcohol for 30 min in a three-litre flask with 10 g of KOH and 6g of Aluminium foil. Distil and collect one litre discarding the first 50 ml. Dissolve 40 g of KOH in 1 litre alcohol. Keep the solution in a stoppered bottle in dark

OR

Crush 40 g KOH in 7 or 8″ mortar. Add 45g granulated calcium oxide and grind mix to powder. From 1 litre of alcohol, add 100 ml to mortar and transfer to flask rinsing mortar with several more portions. Add remaining alcohol to the flask, shake, mix for at least 5 min and invert beaker over neck of the flask. Repeat shaking several times during the day. Next morning filter the solution into clean glass stoppered bottle.

Procedure

- ☆ Weigh 5g of the fat into a 250 or 300ml Erlenmeyer flask. Measure by means of a burette 50 ml of alkaline KOH solution into the flask.
- ☆ Connect the flask with an air condenser and reflux until the fat is completely saponified (about 30 min).
- ☆ Cool and titrate with 0.5N HCl using phenolphthalein as indicator.
- ☆ Perform a blank determination.
- ☆ Subtract the number of ml of 0.5N HCl obtained in the determination of sample from the number of ml of 0.5N HCl equivalent to KOH used in the saponification of the sample.
- ☆ Calculate the saponification number as mg of KOH required to saponify 1g fat.

* Standard 0.5N HCl is made using 0.5N Na_2CO_3 solution and methyl red indicator. Take 0.5N HCl in burette for titration.

Observations

Blank reading

10 ml KOH = 16.5 ml HCl

50 ml KOH = 82.5 ml = say B

Sample, HCl consumes = say S

Calculations

$$\text{Saponification number} = \frac{(B - S) \times 0.0462 \times 1000}{\text{Weight of sample in g}}$$

Method–II: Determination of Saponification Value in a Given Oil Sample

Principle

Saponification value of a meat/fat or oil is the number of mg of KOH required to saponify 1 g of fat. There is an inverse relationship between saponification value and molecular weight. A known weight of sample is reacted with NaOH and unused NaOH is titrated against standard acid solution.

Reagents and Materials

1. Alcoholic Potassium hydroxide solution
2. Hydrochloric Acid (HCl)
3. Phenolphthalein indicator
4. Erlenmeyer flask
5. Air condense

Procedure

- 3ml of oil was weighed in a conical flask and 25 ml of alcoholic KOH was added to it.
- The mixture was refluxed using an air-cooled condenser for 1 h until all fat globules vanished.
- The flask was cooled and titrated against 0.5 N HCl using phenolphthalein as indicator until solution become colorless.
- A blank was run in a similar manner.

Observations

V_2 = Titration of blank

V_1 = Titration of sample

N = Normality of HCl

Calculations

$$\text{Saponification value} = \frac{56.01 \times (V_2 - V_1) \times N}{\text{Weight of sample}}$$

Calculate the theoretical Saponification value of olive oil with the composition 80 per cent oleic acid 9 per cent, linoleic acid, 8 per cent palmitic acid and 1 per cent each of myristic, stearic and arachidonic acid.

Solution

(*a*) Triglyceride Formed by Oleic Acid

$CH_2OCOC_{17}H_{33}$

|

$CH_2OCOC_{17}H_{33}$

|

$CH_2OCOC_{17}H_{33}$

Molecular weight = 88.4

884 requires 168000 mg of KOH

$$\text{Therefore } 0.8 = \frac{168000 \times 0.8}{884} = 152 \text{ mg}$$

(*b*) Triglyceride Formed by Linoleic Acid

$CH_2OCOC_{17}H_{31}$
|
$CH_2OCOC_{17}H_{31}$
|
$CH_2OCOC_{17}H_{31}$

Molecular weight = 878

878 requires 168000 mg of KOH

$$\text{Therefore } 0.9 = \frac{168000 \times 0.9}{878} = 17.22 \text{ mg of KOH}$$

(*c*) Triglyceride Formed by Palmatic Acid

$CH_2OCOC_{15}H_{31}$
|
$CH_2OCOC_{15}H_{31}$
|
$CH_2OCOC_{15}H_{31}$

Molecular weight = 806

806 requires 168000 mg of KOH

$$\text{Therefore } 0.8 = \frac{168000 \times 0.08}{806} = 16.67 \text{ mg of KOH}$$

(*d*) Triglyceride Formed by Myristic Acid

$CH_2OCOC_{13}H_{27}$
|
$CH_2OCOC_{13}H_{27}$
|
$CH_2OCOC_{13}H_{27}$

Molecular weight = 722

722 requires 168000 mg of KOH

$$\text{Therefore } 0.01 = \frac{168000 \times 0.01}{722} = 2.237 \text{ mg of KOH}$$

(*e*) Triglyceride Formed by Stearic Acid

$CH_2OCOC_{17}H_{35}$
|
$CH_2OCOC_{17}H_{35}$
|
$CH_2OCOC_{17}H_{35}$

Molecular weight = 840

840 requires 168000 mg of KOH

$$\text{Therefore } 0.01 = \frac{168000 \times 0.01}{840} = 1.888 \text{ mg of KOH}$$

(*f*) Triglyceride Formed by Arachidonic Acid

$CH_2OCOC_{19}H_{39}$
|
$CH_2OCOC_{19}H_{39}$
|
$CH_2OCOC_{19}H_{39}$

Molecular weight = 974

974 requires 168000 mg of KOH

$$\text{Therefore } 0.01 = \frac{168000 \times 0.01}{974} = 1.725 \text{ mg of KOH}$$

Therefore, Saponification value of olive oil =

$$152 + 17.22 + 16.6 + 2.33 + 1.8 + 1.725 = 191.30$$

21. Peroxide Value

Method–I: Method of US Army Laboratories (Natick) Described by Koniecko (1979)

Reagents and Materials

1. Phenolphthalein indicator (0.2 per cent)
2. Chloroform
3. 0.1N sodium thiocyanate
4. Distilled water
5. Potassium Hydroxide
6. Burette
7. Volumetric flasks
8. Titration flasks
9. Weighing balance
10. Blender/pestle and mortar
11. Whatman filter paper No. 1
12. Hot air oven

Procedure

- ☆ 25 g sample blended for 2 minute with 137 ml of chloroform in presence of ½–1 teaspoonful of anhydrous sodium sulfate.
- ☆ The mix was filtered through Whatman No. 1 filter paper
- ☆ An aliquot (25 ml) of filtered chloroform extract was transferred to a 500 ml conical flask to which 30 ml of glacial acetic acid and 2 ml of saturated potassium iodide solution were added.
- ☆ Let it react for 2 minute with occasional swirling after which 100 ml. of distilled water to stop the reaction and 2 ml of 1 per cent fresh starch solution were added.
- ☆ Flask contents titrated immediately with 0.1N sodium thiocyanate/Sod. thiosulfate to end point (non-aqueous layer turned colorless).
- ☆

  Another 25 ml of the extract were placed in a pre-weighed beaker for estimating the fat weight after vapourization

POV and FFA were calculated and expressed as follows:

$$\text{POV (meq/1000g sample)} = \frac{(0.1 \times \text{ml } 0.1\text{N Sod thiocyanate}^* \times 100)}{\text{Wt. of fat}}$$

* Sod. thiosulfate can be used for estimation of FFA content.)

Method–II: Determination of Peroxide Value in Meat (AOAC, 1975)

Reagents

1. *Acetic acid-chloroform solution*: Mix 3 volumes of acetic acid with 2 volumes of chloroform.
2. *Saturated potassium iodide solution*: Dissolve excess potassium iodide in freshly boiled water. Excess solid must remain. Store in dark. Test daily by adding 0.5 ml to 30 ml of reagent (i) and then add 2 drops 1 per cent starch solution. If solution turns blue, requiring more than 1 drop 0.1 N $Na_2S_2O_3$ to discharge color, prepare fresh solution.
3. *Starch solution*: Mix about 1 g soluble starch with enough cold water to make thin paste, add 100 ml boiling water and boil about 1 minute while stirring.
4. *Sodium thiosulphate standard solution*-0.1 and 0.01 N: dissolve about 25 g $Na_2S_2O_3.5H_2O$ in 1 lit. of water. Boil gently for 5 minutes and transfer while hot to storage bottle previously cleaned with hot chromic acid cleaning solution and rinsed with warm boiled water. Store solution in dark cool place and do not return un-used portion to stock bottle. If solutions less concentrated than 0.1 N are desired, prepare by diluting with boiling water (more dilute solution is less stable and should be prepared just before use).

Standardization: Accordingly weigh 0.20 to 0.25 g $K_2Cr_2O_7$ (dried 2 hours at 100°C) and place in a flask. Dissolve in 80 ml chloride free water containing 2 g potassium iodide. Add, with swirling, 20 ml 1N HCl and immediately place in dark for 10 minutes. Titrate with $Na_2S_2O_3$ solution as prepared above adding starch solution after most of iodine has been consumed.

$$\text{Normality} = \text{g } K_2Cr_2O_7 \times 1000/\text{ml } Na_2S_2O_3 \times 49.032$$

Procedure

- ☆ Weigh 5.0±0.05 g sample in 250 ml flask and add 30 ml acetic acid-chloroform solution and swirl to dissolve.
- ☆ Add 0.5 ml standard KI solution, let it stand with occasional shaking for 1 minute and add 30 ml water.
- ☆ Slowly stir with 0.1 N $Na_2S_2O_3$ with vigorous shaking until yellow is almost gone.
- ☆ Add about 0.5 ml 1 per cent starch solution and continue titration shaking vigorously to reduce all iodine from chloroform layer until blue just disappears. If more than 0.5 ml 0.1 N $Na_2S_2O_3$ is used, repeat determination with 0.1 N $Na_2S_2O_3$.
- ☆ Conduct blank determination daily (must be 0.01 ml 0.1 N $Na_2S_2O_3$).

Calculations

PV (Milli equivalent peroxide/Kg sample) =

$$S \times N \times 1000/\text{g sample}$$

where,

S = ml $Na_2S_2O_3$ (blank corrected)

N = Normality of $Na_2S_2O_3$ solution.

Note: Melt the sample by heating with constant stirring on hot plate or oven (60-70°C) in case of margarine but not more than 70°C for long time.

Method–III: Determination of Peroxide Value of a Given Oil Sample

Principle

The peroxide value of an oil or fat is the amount of peroxide present in it and is expressed as milliequivalents of peroxide present per kg of oil. By peroxide value we may judge the rancidity. Known weight of sample is treated with KI and Iodine liberated is titrated against $Na_2S_2O_3$.

Reagents and Materials

1. Glacial Acetic Acid
2. Chloroform

3. Sodium thiosulphate
4. Conical Flask

Procedure

- ✰ Sample was weighed in a conical flask. 25 ml of solvent (Glacial HAC + $CHCl_3$) was added to it.
- ✰ One ml of KI solution was added and flask was shaked occasionally for two min. 35 ml water was added and titrated against 0.05 N $Na_2S_2O_3$ using starch as indicator.
- ✰ A blank was run in a similar manner.

Observations and Calculations

Weight of sample =

Volume of $Na_2S_2O_3$ used = say x ml

For blank = 0 ml

$$\text{Peroxide value (Meq/kg of oil)} = \frac{1000(\text{x ml}-0)}{\text{Wt of the sample}}$$

Method–IV: Estimation of Peroxide Value of a Given Meat Sample

Peroxide value was estimated as per procedure described in AOAC (1995).

Reagents and Materials

1. Choloroform
2. Acetic acid
3. Potassium Iodide
4. Starch solution
5. 0.1N Sodium thiosulphate
6. Erlenmeyer flask

Procedure

1. Take 5 gm of sample after careful weighing.
2. Transfer the sample into 250 ml glass-stoppered Erlenmeyer flask.

3. Add 30 ml acetic acid-chloroform solution into it and swirled to extract lipid.
4. Add 0.5 ml saturated potassium iodide solution into it.
5. Allowed to stand with occasional shaking for 1 min.
6. Add 30 ml water and 0.5 ml of 0.1 per cent starch solution, Mix it well.
7. The mixed solution was titrated against 0.01 N sodium thiosulphate untill intense blue color disappeared.
8. Always run blank. Blank determinations were conducted and subtracted from sample titration.

$$\text{Peroxide Value (meq/Kg)} = \frac{\text{ml of Sodium thiosulpahte} \times \text{N of Sodium thiosulphate}}{\text{Weight of sample}}$$

22. Free Fatty Acids

Method–I: (Koniecko, 1979)

Reagents and Materials

1. Phenolphthalein indicator (0.2 per cent)
2. Chloroform
3. Distilled water
4. Potassium Hydroxide
5. Burette
6. Volumetric flasks
7. Titration flasks
8. Weighing balance
9. Blender/pestle and mortar
10. Whatman filter paper No. 1
11. Hot air oven
12. Anhydrous sodium sulfate

Procedure

- ✰ 25 g sample blended for 2 minute with 137 ml of chloroform in presence of ½–1 teaspoonful of anhydrous sodium sulfate.

- ☆ The mix was filtered through Whatman No. 1 filter paper
- ☆ An aliquot of 25 ml extract was transferred to 125 ml conical flask and10 drops of 0.2 per cent phenolphthalein indicator were added to it.
- ☆ Titrated with 0.1N 90 per cent alcoholic potassium hydroxide to the end point (pink color).
- ☆ Another 25 ml of the extract were placed in a pre-weighed beaker for estimating the fat weight after vapourization and removal of chloroform at 80°C in a drying oven.

FFA were calculated and expressed as follows:

$$\text{FFA (\% Oleic acid)} = \frac{(0.1 \times \text{ml } 0.1\text{N alcoholic KOH} \times 0.282 \times 100)}{\text{Wt. of fat}}$$

Method–II

The free fatty acid contents were determined by the procedure as described by ISI (1966).

Reagents and Materials

1. 0.1N Sodium Hydroxide
2. Hot water bath
3. Burette
4. Conical flask

Procedure

- ☆ Accurately 10 g of the sample was weighed in conical flask.
- ☆ In a second flask, 50 ml of alcohol was brought to the boiling point and while still above 70°C neutralized it with 0.1 N sodium hydroxide.
- ☆ Poured the neutralized alcohol on sample in the flask and mixed the contents of the flask.
- ☆ Brought them to boil and while it was still hot, titrated with 0.1 N sodium hydroxide, shaken vigorously during the titration.

- ☆ The end point of titration was reached when the addition of a single drop produced a slight but definite color change persisted for at least 15 seconds.

Free fatty acids (per cent) in the samples calculated as oleic acid.

$$\text{Fat fatty acids (per cent)} = \frac{2.82\ T}{W}$$

where,

T = Volume of 0.1 N alkali required for titration in ml, and

W = Weight of sample taken.

Method–III

Reagents and Materials

1. Ether Extract
2. Phenolphthalein indicator
3. 0.01N KOH
4. Conical flask
5. Burette

Procedure

- ☆ Fat in the samples of meat sample was extracted by Soxhlet method (AOAC, 1995).
- ☆ After recovering fat, 30 ml of neutralised alcohol and 1 ml of phenolphthalein indicator added.
- ☆ Titrated the contents against 0.01 N KOH solution.

Per cent free fatty acid was calculated as follows:

$$\text{\% free fatty acids (as oleic acid)} = \frac{\text{Volume of N/100 KOH solution used} \times 28.2 \times .1}{\text{Weight of sample}}$$

Method–IV: Determination of Acid Value or Free Fatty Acid Content

Principle

Acid value is defined as the number of mg of KOH required to neutralize the free fatty acids present in one gram of fat or oil. It gives an idea about storage and quality of oil and fats. The formulae used are under mentioned.

$$\text{Free fatty acid (\% oleic acid)} = \frac{\text{Volume of KOH} \times N \times 28.2}{\text{Weight of sample}}$$

$$\text{Free fatty acid (\% of lauric acid)} = \frac{\text{Volume of KOH} \times N \times 20.2}{\text{Weight of sample}}$$

$$\text{Free fatty acid (\% palmitic acid)} = \frac{\text{Volume of KOH} \times N \times 25.6}{\text{Weight of sample}}$$

Reagents and Materials

1. KOH (N/10)
2. Phenolphthalein indicator
3. Neutral alcohol
4. Burette
5. Conical flask

Procedure

- ☆ 5g of oil was weighed in a conical flask
- ☆ 2.5 ml of neutral alcohol was added and shake well
- ☆ It was titrated against KOH using phenolphthalein as an indicator

Observations and Calculations

Weight of sample =

Volume of KOH used = say × ml

Normality of KOH = 0.1

$$\text{Acid value (per cent)} = \frac{56.1 \times \times \text{ ml} \times 0.1}{\text{Wt. of the sample}}$$

$$\text{Free fatty acid as oleic acid} = \frac{28.2 \times \times \text{ ml} \times 0.1}{\text{Wt. of the sample}}$$

Method–V: Estimation of Free Fatty Acid

Free fatty acid value was determined by modified AOCS method (Koniecko, 1979).

Reagents and Materials

1. Petroleum ether
2. Pestle and mortar
3. Measuring cylinder
4. Filter paper No. 42
5. Petri plate
6. Hot air oven
7. Ethyl alcohol
8. Phenolphthalein indicator
9. 0.1N NaOH

Procedure

- ☆ Take 15 g of meat sample
- ☆ Blend it with the help of pestle and mortar
- ☆ Dissolve the meat in 150 ml of petroleum ether,
- ☆ Shake well and filtered through Whatman's filter paper No. 42
- ☆ Collect about 90 ml of filtrate.
- ☆ Then take two parts of filtrate measuring 20 ml each.
- ☆ One part was taken in preweighed Petri plate for different test groups.
- ☆ Dry it in hot air oven at 100±2° C and fat weight was calculated.
- ☆ Another part of 20 ml filtrate of different test groups was taken in different beakers.
- ☆ Take 10 ml neutral ethyl alcohol was added to filtrate of each group. Then it was shaken well for proper dissolution of fat, two drops of phenolphthalein indicator (1 per cent) was added to the filtrate of beaker and was titrated against 0.1N NaOH till the end point was obtained. The free fatty acid content of the sample was calculated as:

$$\% \text{ FFA as Oleic Acid} = \frac{\text{N} \times \text{ml of NaOH used in titration} \times 0.282}{\text{Fat weight}} \times 100$$

23. Determination of Iodine Value of the Given Sample

Principle

Iodine value is defined as the number of grams of Iodine absorbed by 100 gm of fat. It is a measure of unstauration. Higher iodine is the indication of higher unsaturation and vice versa. Known weight of the sample is reacted with iodine and the excess of iodine is determined by titration against $Na_2S_2O_3$ solution.

Reagents and Materials

1. Glacial acetic acid
2. Potassium Iodide solution (15 per cent)
3. Carbon tetrachloride
4. 0.1 N Sodium Thiosulphate solution
5. Starch solution
6. Iodine solution
7. Test tubes
8. Beakers

Procedure

- ✰ 10 g of Peanut oil sample was weighed in a conical flask. 20 ml of Carbon tetra chloride and 25 ml of Iodine solution was added. It was kept in dark for 30 min.
- ✰ 20 ml of 15 per cent KI solution was added and 50ml of recently boiled cooled water was added.
- ✰ It was titrated against 0.1 N $Na_2S_2O_3$ with starch as an indicator.
- ✰ The end point is distinguished by disappearance of blue color.
- ✰ A blank was run in a similar way.

Observations

w = Sample weight in g

A = Sample titration reading in ml

B = Blank reading in ml

Calculations

$$\text{Iodine Value} = \frac{(B - A) \times \text{Normality of } Na_2S_2O_3 \times 126.9 \times 100}{W \times 1000}$$

Precautions

- ☆ The conical flask must be stoppered and placed immediately after adding reagent in dark.
- ☆ The time must be strictly followed and should be homogenous for sample as well as blank.

24. Extractable Protein

Total protein in the muscle homogenates was determined by the Kjeldahl nitrogen method (AOAC, 1984). The procedure for determining extractable protein (EP) was a modification of the procedure used by Saffle and Galbreath (1964).

Reagents and Materials

1. Sodium chloride solution
2. Refrigerated centrifuge
3. Homogenizer/blender
4. Spatula

Procedure

- ☆ Two 5g samples of were diluted with 20 ml of an appropriate unbuffered NaCl solution to attain a final concentration of 1.0 M NaCl.
- ☆ Samples were homogenized for 3 sec, allowed to stand for 3 min, and homogenized again for 3sec.
- ☆ The slurry was centrifuged at 16000 g for 12 min. After centrifugation the fat layer floating on top was gently removed with a stainless steel spatula.
- ☆ The supernatant was decanted and centrifuged at 16000 g for 12 min.

- ☆ The second supernatant was analyzed for protein concentration by the Kjeldahl nitrogen method.
- ☆ EP was expressed as percent of the total meat protein

25. Protein Fractions

Muscle proteins were fractioned on the basis of their solubility in water/salt solution using the method of Kang and Rice (1970).

(*i*) Water Extractable Protein (WEP)

Method–I

Reagents and Materials

1. Refrigerated centrifuge
2. Homogenizer/blender
3. Chilled distilled water
4. Reagents of Lowry's method of protein estimation

Procedure

- ☆ 4 g of meat sample was homogenized with 30 ml of chilled distilled water in a Ultra Turrex mixer for 2 minute.
- ☆ Transfer to a 100 ml conical flask and kept overnight at 4°C.
- ☆ The slurry was centrifuged at 4°C in a refrigerated centrifuge at 5000 rpm for 5 minute and the supernatant was collected.
- ☆ The residue was re-extracted with 10 ml of chilled distilled and centrifuged as above.
- ☆ The supernatants were pooled together and volume was made up to 50 ml with chilled distilled water.
- ☆ Protein content was determined by the method of Lowry *et al.* (1951).
- ☆ Water extractable protein was expressed as per cent.

Method–II: Water soluble Fraction (Sarcoplasmic Proteins)

Reagents and Materials

1. Chilled distilled water

2. Refrigerated centrifuge
3. Homogenizer/blender
4. Micro Kjeldahl Assembly and its reagents

Procedure

- ✰ Accordingly weigh 20 g minced meat and disperse in a warring blender for two minutes by adding 25-30 ml cooled distilled water.
- ✰ Transfer the homogenate quantitatively into a 200 ml volumetric flask and make up the volume up to the mark.
- ✰ Keep the homogenate at 4-6°C overnight and then centrifuge for 15 minutes at 3000 rpm in a refrigeration centrifuge at 4°C.
- ✰ Estimate the protein in the supernatant by micro-Kjeldahl method (AOAC, 1975).

(*ii*) Salt Extractable Protein (SEP)

Method–I

Reagents and Materials

1. Chilled 0.67M NaCl solution
2. Homogenizer
3. Refrigerated centrifuge
4. Reagents of Lowry's method of protein estimation

Procedure

- ✰ Salt extractable protein were obtained by homogenizing the residue (remaining after the extraction of water soluble proteins) with 30 ml of chilled 0.67 M NaCl solution in a Ultra Turrex homogenizer for 2 minute and left overnight at 4°C.
- ✰ The slurry centrifuged in a refrigerated centrifuge at 5000 rpm for 5 minutes and the supernatant collected.
- ✰ Residue was re-extracted with 10 ml of 0.67 M NaCl and centrifuged.

- ☆ The supernatants were pooled and made up to 50 ml volume.
- ☆ Method of Lowry *et al.* (1951) was followed for protein estimation and result was expressed as per cent.

Method–II: Salt Extractable Proteins

Slightly modified method of Knipe *et al.* (1985) was followed.

Reagents and Materials

1. 0.6 M NaCl
2. Homogenizer
3. Refrigerated centrifuge
4. Mechanical stirrer
5. SS spatula

Procedure

- ☆ Finely minced 5 g meat sample was homogenized with 15 ml of chilled 0.6M NaCl for 1 min using pestle and mortar.
- ☆ Then 10 ml of chilled 0.6 M NaCl was added and homogenized for another 1 min.
- ☆ This homogenate was quantitatively transferred with two rinsing to 65 ml polycarbonate centrifuge tubes with press cap and the final volume was made to 50 ml.
- ☆ The samples were stirred on a mechanical shaker for 2 min, and centrifuged at 5500 rpm for 15 min in a centrifuge.
- ☆ After centrifugation, the fat layer floating on the surface was gently moved to one side with a stainless steel spatula and 1 ml aliquot in duplicate was drawn from the clear solubilized protein solution.
- ☆ To each 1 ml of solution, 5 ml Biuret reagent was added.
- ☆ This mixture was stirred and allowed to stand for 15 min for optimum color development.
- ☆ Optical density was determined with a spectrophotometer at 540 nm and converted by using bovine serum albumin (BSA) standard curve to mg protein per ml solution.
- ☆ Salt extractable protein was expressed as g per 100g meat.

(*iii*) Non-Extractable Protein Fraction (Stromal and Denatured Proteins) (NEP)

The residue remaining after extraction of salt extractable proteins was used for determination of non-extractable proteins. Procedure followed was same as described for total protein and expressed as per cent.

26. Available Lysine

Method–I

The method of Carpenter (1960) as modified by Booth (1971) was followed to determine the available lysine content in meat.

Reagents and Materials

1. 1M Na_2CO_3 solution
2. 1-fluro 2, 4- dinitrobenzene solution
3. Ethanol
4. 0.1M HCl
5. Methoxy carbonyl chloride
6. Spectrophotometer
7. Hot water bath
8. Measuring cylinders
9. Weighing balance

Procedure

- ✰ About 1 g of defatted and dried sample was taken in a round bottom flask and 10 ml of 1M Na_2CO_3 solution was added and carefully left to stand until the sample was wetted
- ✰ 15 ml of the 1-fluro 2, 4- dinitrobenzene solution (about 0.4 ml of FDNB in 15 ml of ethanol) was added to the flask and shaken gently for 2 hrs.
- ✰ The ethanol was evaporated on a boiling water bath.
- ✰ When the mixture cooled, 30 ml of 0.1M HCl were added to neutralize the sodium bicarbonate.
- ✰ The mixture was refluxed gently for 16 hrs.
- ✰ The contents were filtered hot through Whatman filter paper No. 1 and the volume was made upto 250 ml.

- ☆ 2 ml of the filtrate was pipetted into two stoppered test tubes A and B separately.
- ☆ The contents of test tube B were extracted with about 5 ml of peroxide free diethyl ether.
- ☆ The tube was placed in hot water (about 80°C) until effervescence from the residual ether ceased, and allowed to cool.
- ☆ A drop of phenolphthalein solution was added followed by NaOH solution from a dropping pipette until the fist pink color appeared
- ☆ 2 ml of carbonate buffer (pH 8.5) was added.
- ☆ Five drops (about 0.01 ml) of Methoxy carbonyl chloride (M.C.C.) was then added under the fume hood.
- ☆ The tube was firmly stoppered and shaken vigorously
- ☆ After 8 min. 0.75 ml of conc. HCl was added.
- ☆ The solution was extracted four times with ether and the tube kept in hot water for 5 min.
- ☆ The tube was cooled and the contents made upto 10 ml with water.
- ☆ During the pauses between manipulations of tube B, tube A, was extracted three times with peroxide free diethyl ether.
- ☆ Residual ether was removed as described above and the contents made upto 10 ml with 1 M HCl.
- ☆ The absorbance of both A and B were read at 435 nm against water as a blank.
- ☆ The absorbance attributable to DNP-lysine was obtained by subtracting reading B from reading A.
- ☆ Mono-ε-N-dinitrophenyl–lysine hydrochloride monohydrate (DNP-L) was used as standard.
- ☆ The concentration was read by the following regression equation constructed from the reference values

$$Y = 0.02134+2.0179\ X$$

Method–II: (James and Ryley, 1986)

The basic method of Hall *et al*. (1973) was used as the starting point of the study.

Reagents and Materials

1. Conc. HCl
2. Standard lysine stock solution*
3. Working solution of lysine (100-500µg/ml)
4. $NaHCO_3$
5. 10g litre^{-1} TNBS solution
6. 1.0 M $NaHCl_3$
7. Hot water bath
8. Spectrophotometer

* A standard lysine stock solution (0.6–0.7 g D,L.-Lysine hydrochloride/100 ml boiled out distilled water containing 0.1 ml Octan-2.0 l was stored at 4°C for up to 4 weeks.

Procedure

- ☆ A finely divided sample containing 20–60 mg of nitrogen was homogenized for 5 min in 80 ml of agar;
- ☆ 1g of the suspension was treated with 1.25 ml of 1.0 M $NaHCl_3$ followed by 2.5 ml of 10g litre^{-1} TNBS solution in a water bath at 40°C for 75 min.
- ☆ The mixture was then acidified with 7.5 ml concentrated HCl and hydrolyzed in a boiling water bath for 2 hr.
- ☆ Blanks were prepared by acidifying the sample prior to the incubation period, a procedure which inhibited the trinitrophenylation reaction.
- ☆ At the end of the hydrolysis period, the samples and blanks were filtered and the filtrate made up to 25 ml; 10 ml aliquots were extracted twice with ether.
- ☆ The final traces of ether were removed by heating on a water bath at 60°C and the aqueous residue made up to 25 ml.
- ☆ The absorbance was read at 415 nm and the absorbance of the samples connected by subtraction of the samples connected by subtraction of the absorbance of the blank.
- ☆ The lysine concentration was interpolated from a calibration graph prepared as follows : a standard stock solution (0.6–0.7 g D,L.-Lysine hydrochloride/100 ml

boiled out distilled water containing 0.1 ml Octan-2.0 l was stored at 4°C for up to 4 weeks.

- A diluted stock (10 ml per 100 ml) was used to prepare a series of standards by pipetting 0.2, 0.4, 0.6, 0.8 and 1.0 ml (100–500 μg of lysine and making up to 1 ml with water, treating with $NaHCO_3$ and TNBS as for the samples and followed by hydrolysis for 2 hr.

Advantages

- It may be applied to a much wider range of sample.
- It gives very similar results to Carpenter's method to pure proteins and to the Silcock method in the presence of carbohydrates.
- E-TNP lysine hydrolyzed for 90 minute or D,L–lysine taken through the trinitrophenylation procedure followed by hydrolysis for 120 minute are the most suitable standards.
- The procedure can be completed in under 6 hour and does not require the use of expensive equipment or hazardous chemicals.
- This modified method compared well with the FDNB methods and is a fast, accurate safe, and suitable method for routine analysis.

27. Non-Protein Nitrogen

Method–I

NPN content of is determined according to the method described by Rowland (1938) with some modifications.

Reagents and Materials

1. 15 per cent Trichloro Acetic Acid
2. Whatman No. 40 filter paper
3. Pestle and mortar
4. Conical flasks

Procedure

- Ten grams of meat sample is transferred to a porcelain pestle mortar, made into paste by adding 15 per cent TCA.

- ☆ The contents were mixed thoroughly and transferred quantitatively to a volumetric flask.
- ☆ The volume was made up to 100 ml by adding TCA (15 per cent).
- ☆ After thoroughly mixing, the contents were filtered through Whatman filter paper No. 40.
- ☆ Twenty-five ml of the filtrate representing 2.5 g of sample was used for nitrogen determination; Per cent non-protein nitrogen is calculated.

Method–II: Non-Protein Nitrogen

NPN-content of meat products was analyzed as per the procedure of Hegarty *et al*. (1963)

Reagents and Materials

1. 10 per cent Trichloro Acetic Acid
2. Whatman No.1 filter paper
3. MicroKjeldahl Assembly and reagents

Procedure

- ☆ 15 ml of water soluble (sarcoplasmic) extract was mixed with 5 ml of 10 per cent trichloroacetic acid (TCA)
- ☆ After 15 min., it was filtered through Whatman No. 1 filter paper.
- ☆ The filtrate was analyzed for non protein nitrogen by MicroKjeldahl method (A.O.A.C.1980).

Note: To obtain water soluble extract, accurately weigh 20g minced meat and disperse in a warring blender for 2 min by adding 25-30 ml cooled distilled water. Transfer the homogenate quantitatively into a 200 ml vol. flask and make up the volume upto the mark. Keep the homogenate at 4-6°C overnight and then centrifuge at 4°C. Use the supernatant as w.s. ext.

Method–III: Non-Protein Nitrogen

Reagents and Materials

1. 30 per cent Potassium hydroxide
2. Centrifuge

3. 0.6N perchloric acid
4. MicroKjeldahl assembly and reagents

Procedure

- ✰ Take 5 g meat homogenized in 25 ml 0.6N perchloric acid for 3 min at maximum speed.
- ✰ Mixture centrifuged (10 min at 7000g), residue washed with 25 ml of 0.6N perchloric acid, filtered along with supernatant.
- ✰ Filtrate neutralised with 30 per cent (w/v) KOH sol., again filtered and diluted to 100 ml distilled water.
- ✰ Total NPN determined by modified Kjeldahl method (AOAC, 1984).
- ✰ NPN expressed as mg/100g dry matter.

28. Free Amino Acids

Glutathione and glutamine were found to decrease in the meat during storage, while most other amino-acids, notably glutamic acid and tryptophan increased. Increased glutamic acid related to the microbial nos. in the meat, increased tryptophan seemed directly related to the time and temperature of storage. (Gardener and Stewart,1966)

Initial increase urea-N followed by a residual disappearance of this compound from stored meats. Ammonia-N did not rise significantly until the stage of incipient spoilage had been reached.

The method of Rosen (1957) was followed to determination the free amino acids in NPN fraction by ninhydrin reagent.

Reagents and Materials

1. 10 per cent Trichloro acetic acid
2. Ninhydrin reagent
3. Centrifuge
4. Spectrophotometer

Procedure

- ✰ To 10 ml of water soluble (sarcoplasmic) protein fraction 5 ml of 10 per cent TCA was added for precipitation of the true proteins.

- ☆ The contents were centrifuged for 15 min. and the supernatant was carefully decanted.
- ☆ 2 ml of the supernatant was diluted to 100ml in a volumetric flask.
- ☆ From this, 2 ml of solution was taken in a test tube and 1 ml of ninhydrin reagent was added
- ☆ The above test tube was heated in a water bath for 10 min. and cooled under running tap water.
- ☆ The absorbancy was measured at 530 nm in a spectrophotometer.
- ☆ The concentration was read by following regression equation constructed from the reference values.

$$Y = 0.002+3.44\ X$$

Reference values: 65.5 mg of Leucine was dissolved in 100 ml of distilled water. From this solution 0.5, 1.0, 1.5, 2.0 and 2.5 ml were taken in volumetric flask and the volume was made up to 100 ml with water. Two ml each from these concentrations were treated as above.

29. Determination of Some Problematic Amino Acids

Tryptophan

Tryptophan is an essential amino acid and its determination in the free form is easy by any of the previously proposed methods. However, the determination of protein tryptophan is difficult since this amino acid is destroyed during acid hydrolysis, especially in the presence of carbohydrates. The first proposal would be the addition of some protective agents against oxidation (e.g, tryptamine, mercaptoethanol, thioglycolic, mercaptopropionic, or thio-or dithio-dipropionic acids). The second alternative would be the use of organic acids to hydrolyze the protein, such as mercaptoethanesulfonic acid or 4 N methanesulfonic acid, which give good results for samples with less than 20 per cent of sugar. Perhaps alkaline hydrolysis is the most advisable method for a better tryptophan recovery even in the presence of high amounts of carbohydrates. As a last alternative, enzymatic hydrolysis has been used for the determination of tryptohan in infant formulas and medical nutritional products, but the technique is complex.

Cystine, Cysteine and Methionine

The importance of these sulphur amino acids is due to their high reactivity and reducing power. Methionine is an essential amino acid, while cyst(e)ine is essential in premature infants. These amino acids are especially important in the animal feed industry because of the high requirements in the growth of hair and feather.

The analysis of methionine and cyst(e)tine is problematic because the sulphur-containing side chains are easily oxidized at a variable rate. Many methodologies have been proposed including chromatography or colorimetry, but problems arise when an hydrolyzing step is necessary for the analysis. Indeed cyst(e)tine and methionine are partially oxidized during acid hydrolysis; yielding several adducts, cystine, cysteine, cystein sulfinic acid, cysteic acid methionine, methionine sulfoxide and methionine sulfone which makes analysis difficult. The normal procedure for overcomeing this problem is to convert all cyst(e)tine and methionine into cysteic acid and methionine sulfonic acid by sample treatment with performic acid and analyzing them in this form by any of the methods previously described. The problem is the poor retention of the derivatized cysteic acid in any of the described chromatographic methods (cation-exchange or RP-HPLC methods). The use of alkylating agents to stabilized the cysteine before hydrolysis is a valid alternative since they offer derivatives with good chromatographic behaviour. The determination of methionine is not so problematic because less losses occur during acid hydrolysis (the addition of a protective against oxidation and absence of oxygen is enough for a good recovery), except in samples with a high content of carbohydrates where performance oxidation is advisable.

30. Total Volatile Nitrogen (TVN) Pearson (1968a)

In meat, volatile nitrogen mainly consists of ammonia. As ammonia production due to deamination of amino acids increases during storage spoilage, its determination represents a simple method of following the courses of deterioration of the lean meat (Pearson 1968a). Most beef can be considered to be acceptable if the TVN figure does not exceed 16.5 mg N/100g.TVN in prepared minced meat samples were determined in accordance with the procedure described by Pearson (1968c) following micro-diffusion technique with slight modification.

Reagents and Materials

1. Trichloroacetic Acid
2. Potassium carbonate
3. Sulphuric Acid (H_2SO_4)
4. Pestle and mortar
5. Whatman Filter Paper No. 40
6. Conway micro-difusion unit
7. Boric Acid reagent

☆ *Preparation of Boric Acid Reagent*: 5 g boric acid was stirred with 100 ml of ethyl alcohol (95 per cent) in a 500 ml volume flask and 350 ml distilled water was added. Acid was properly dissolved and 5 ml of indicator (0.66 per cent methyl red and 0.033 per cent bromocresol green in alcohol) was added. A faint reddish color was developed by adding 40 per cent NaOH solution. The final volume was made up to the mark with alcohol.

Procedure

☆ 100g minced meat was triturated thoroughly in a china-clay pestle mortar with 5 g Trichloroacetic acid (TCA)

☆ Covered with Al. foil to prevent moisture loss and kept at room temperature for 30 minute.

☆ Then filtered through muslin cloth. Filtrate then obtained was re-filtered through a Whatman filter paper No. 40.

☆ Thoroughly cleaned, dried Conway micro-difusion unit was taken and 2 ml of boric acid reagent was added in its centre compartment.

☆ One ml of meat filtrate (in last phase of storage 0.5 ml meat filtrate) was accurately pipetted into the outer compartment.

☆ Cover lid was then put in such a way so that only a small portion of outer compartment sufficient to insert the pipette remained open and 1 ml of saturated K_2CO_3 solution was then added through the gap.

☆ Lid was immediately closed without leaving any space. Dish was rotated manually to ensure proper mixing of meat extract with sat. K_2CO_3 solution and then incubated at 37°C for 3-4 hour. During incubation dishes were again rotated 2-3 times.

- After incubation boric acid solution in the centre compartment (faint reddish color of boric acid changed to green color) was titrated with 0.02N H_2SO_4.
- Diffusion was carried out in duplicate along with a blank.

TVN content was calculated by using the formula:

mg TVN/100 ml meat juice =

Reading of burette × Normality of acid used
for titration × 14 × 100

31. Determination of Hydroxyproline in Muscle

Method I: Using HPLC with 9-fluroenylmethyl chlorformate-1-aminoadamantane (FMOC-ADAM)

Collagen is ubiquitous animal protein that imparts important structural and tensile strength properties to muscle as well as textural characteristics to meat and meat products. The unique amino acid composition of collagen, which contains considerable amounts of 4-hydroxyproline (HYP), affords a means by which collagen content of a sample may be estimated.

HPLC can be used as rapid, sensitive and reproducible method for the determination of 4-Hydroxyproline (HYP). It involves the usage of 9-fluroenylmethyl chlorformate-1-aminoadamanatane (FMOC-ADAM) for determination of HYP in meat and meat products (Betner and Foldi, 1988)

In the HPLC method, a standard curve is established using the peak height of HYP solutions of known concentrations (0-10mg/L). The HYP concentration of unknown meat sample is determined from this standard curve.

Reagents and Materials

1. HPLC equipped with a single piston reciprocating pump to which two propotionating valves are mounted.
2. Injector fitted with 10 µl sample loop.

3. A Perkin-Elmer model 650-10S fluorescence spectrophotometer to monitor the column effluent with excitation emission maxima at 260 nm and 310 nm, respectively and both slits set at 10nm.
4. Analytical column (4.6 × 150mm) packed with 5µm Spherisorb C-8 stationary phase
5. Altech guard column (4.6 × 30 mm) packed with identical material.
6. Hewlett Packard Model 3390A integrator
7. HPLC grade acetonitrile
8. Ortho-phthaldialdehyde (OPA): 2- mercaptoethanol (2ME) reagent (50 mg OPA and 26µL 2ME/ml)
9. 2- mercaptoethanol (2ME)
10. FMOC (4mM FMOC reagent dissolved in acetone)
11. Iodacetamide reagent (140 mg/ml)
12. ADAM (2mM ADAM reagent dissolved in acetonitrile)
13. p-DABA
14. 4-hydroxyproline
15. Choramine T
16. Elution buffer : A. It consist of 100mM phosphoric acid.
    - ☆ B. 10 0 per cent acetonitrile
    - ☆ C. 0.8M borate reaction buffer and pH adjusted to 9.5 with sodium hydroxide.

Procedure

(a) Sample Preparation and Derivatization

- ☆ All muscle and tendon samples are frozen in liquid nitrogen and powdered.
- ☆ Sample hydrolysis is performed by refluxing with 6N HCl at 121°C and 15 psi for 16 h.
- ☆ After mixing with 1-2 g decolorizing carbon, the hydrolysates are filtered through Whatman filter paper No. 1, evaporated at 70°C, adjusted to pH 6.5 with 1N NaOH, refiltered through Whatman filter paper No. 1 and brought to volume.

- ✰ A 50µL aliquot of the hydrolysate is then mixed with 850 µL deionized water and 100 µL borate reaction buffer. A 100 µL aliquot of OPA-2ME reagent is added and allowed to react for 30 s.
- ✰ Iodacetamide reagent (100µL) is then added with another reaction interval of 30 s.
- ✰ FMOC reagent (200µL) is added and immediately upon mixing, 1 ml ADAM reagent is added to react with the excess FMOC.

(b) Chromatography

- ✰ The elution programme followed for the elution of HYP is:
- ✰ The first 37 per cent solvent B for 1.5 min., a gradient of 37-42 per cent B is maintained over the next 1.5 min., a column flush of 100 per cent B for 1 min., followed by column re-equilibration with 37 per cent B for 3.5 min.
- ✰ The elution programme is total for 7.5 min. at a flow rate of 2 ml/min.
- ✰ The injection volume is 10 µl.
- ✰ A standard curve is established using the peak height of HYP solutions of known concentrations (0.10 mg/litre). The HYP concentration of unknown solutions is determined from this standard curve, which is prepared daily.

Determination of 4-Hydroxyproline and 5-Hydroxyproline

4- Hydroxyproline and 5- Hydroxyproline are specific amino acids inside the primary structure of collagen is a low-quality protein, since it is poor in essential amino acids content, and thus, its presence in foods decreases their value. These amino acids have been used as an index of the collagen content .

The analysis of these amino acids is achieved through any of the described methods, RP-HPLC being the most used, but taking into account that 4- Hydroxyproline is secondary amino acid that does not react with some derivative reagents such as OPA unless some precautions are considered. In addition to HPLC methods specific colorimetric methods and CG-MS of the N-TFA propyl ester derivatives have also been proposed.

Method–II

Determine the hydroxyproline content of muscle first and convert it into collagen content by multiplying it with 7.25 (Stegman and Stalder, 1967; ISO, 1974; Goll *et al.*, 1963).

Reagents and Materials

1. *Oxidation reagent*: 1.41 g of chloramin T, 10 ml of water, 10 ml of propanol-1 and 80 ml of buffer as in serial (ii).
2. Buffer-50 g citric acid monohydrate, 12 ml of acetic acid, 120 g sodium acetate tetrahydrate and 34 g NaOH, diluted to 1 litre with water. Mix this solution with 200 ml of water and 300 ml of propanol-1.
3. *Color reagent*: Add 65 ml propanol-2 slowly to 35 ml perchloric acid containing 10 g 4-dimethyl amino-benzaldehyde.

Procedure

- ☆ Accurately weigh 4.0 g of thoroughly minced meat into 200 ml boiling flask equipped with water cooled condensor.
- ☆ Add 100 ml of 6 N HCl and continue hydrolysis on a water bath for 16 hours.
- ☆ Filter the hot hydrolysate through a sintered glass funnel (200 ml G 4). Wash the flask three times with hot 6 N HCl and add to the hydrolysate for filtration.
- ☆ Dilute the hydrolysate with water to make up to 200 ml. Take 5 ml of the hydrolysate in beaker and neutralize with 10 per cent NaOH initially followed by 1 N NaOH for adjusting to pH 6.0. Make up the volume of the neutralized hydrolysate to 100 ml with distilled water.
- ☆ Take 4.0 ml of hydrolysate in a test tube and add 2.0 ml of oxidation reagent. Allow the content to be agitated on a shaker and let it stand for 20 minutes at room temperature. Add 2.0 ml of freshly prepared color reagent to the test tube and mix thoroughly. Transfer the test tube quickly to a water bath at 60°C for 15 minutes.

- ☆ Cool the test tube under tap water for 3 minutes.
- ☆ Measure the absorbance in a spectrophotometer at 560 nm. Calculate the concentration from the regression equation constructed form the reference values.

Reference values for hydroxyproline–To 100 mg of L-Hydroxyproline in water add 1 drop of 6 N HCl and make up the volume with distilled water to 100 ml. Keep this in a refrigerator as stock solution. On the day of use, take 5 serial dilutions containing 0.5, 1.0, 1.5, 2.0 and 2.5 µg/ml and treat with oxidizing and coloring agents as described above. Read the color developed at 560 nm for calculating the regression equation.

Method–III

Total Collagen content

Amount of collagen in the sample was calculated by estimating hydroxyproline content according to the procedure of Neumann and Logan (1950).

- ☆ Three gm of sample was hydrolyzed with 50 ml of 6N HCl for 16 hrs at 110°C under reflux.
- ☆ The hydrolysate was diluted to 100 ml.
- ☆ Suitable aliquot was neutralized to pH 7.0, filtered and made upto 100 ml by diluting with distilled water so as to bring the hydroxyproline (HYP) conc. within the range of 1 to 4 µg/ml.
- ☆ To 1ml sample aliquot in a test tube, 1 ml each of 0.01 M copper sulfate, 2.5 N NaOH, and 6 per cent H_2O_2 over added in succession.
- ☆ The solutions were mixed and kept at room temperature for 5 min. with occasional shaking.
- ☆ The tubes were then placed in a water bath at 80°C for 5 min with frequent vigorous shaking and were chilled in ice.
- ☆ 4 ml of 3 N H_2SO_4 and 2 ml of 5 per cent *p*-dimethyl amino benzaldeleyde in n-propanol were added and mixed thoroughly.
- ☆ The tubes were placed in a water bath at 70 ° C for 16 min.

- ☆ The tubes were then cooled under tap water
- ☆ O.D. was read at 540 nm in a spectrophotometer. Suitable standard and distilled water blank was carried through simultaneously.

The factor of 7.25 was used to convert Hydroxy Proline into collagen content and expressed as mg/g wet tissue.

Soluble Collagen

Measured using the procedures of Hill (1966) method with some modifications.

Reagents and Materials

1. Ringer's solution
2. Conc. Hydrochloric Acid
3. Hot water bath shaker
4. Whatman No. 2 filter paper

Procedure

- ☆ Approximately 4g samples were weighed in duplicate suspended in 12 ml of one-quarter-strength Ringer's solution.
- ☆ The mixture was heated for 70 minute at 77°C in a shaking water bath.
- ☆ The tubes were centrifuged, the supernatant was decanted and the residue was washed thoroughly with 6 ml of the Ringer's solution. The supernatants were combined and brought to volume.
- ☆ A filtration step was added using Whatman No. 2 to remove any particulate matter from the supernatant.
- ☆ Equal volumes of concentrated HCl and supernatant were combined in hydrolysis tubes (bringing the final HCl concentration to 6N).
- ☆ The samples were hydrolysed and assayed for hydroxy protein in the same manner as per total collagen analysis.
- ☆ Soluble collagen was expressed as mg of collagen per gm of wet tissue and as percentage of soluble collagen, which

was calculated as soluble collagen (mg/g wet tissue) divided by total collagen (mg/g wet tissue).

Insoluble Collagen

It was calculated as the difference between total collagen and soluble collagen.

Method–IV

Collagen content

Total Collagen content of the raw meat samples are estimated by determining the hydroxyproline content, which represents around 13-14 per cent of the collagen content of the meat. Hydroxyproline (HP) content of the meat sample is determined based on the procedure of Neuman and Logan (1950), with some modifications.

Reagents and Materials

1. NaOH solution
2. Copper sulphate
3. Hydrogen peroxide
4. 3N H_2SO_4
5. 5 per cent DMEB (4-dimethylaminobenzaldehyde)
6. n-propanol
7. Standard hydroxyl propanol
8. Hot water bath
9. Spectrophotometer

Procedure

1. Take 2 g of meat sample and hydrolyze it with 40 ml of 6 N HCl at 105°C for 18 hrs.
2. The hydrolyzed sample is filtered and the volume is adjusted to 50 ml with distilled water.
3. Suitable aliquot (25 ml) is taken, pH was adjusted to 7.0 with 40 per cent NaOH and the volume is made to 50 ml.
4. From this, 1 ml aliquot was taken in a test tube.
5. Add 1 ml each of 0.01M copper sulphate, 2.5 N NaOH and 6 per cent H_2O_2

6. In blank, instead of 1 ml aliquot, 1 ml of distilled water is taken.
7. After thorough mixing, the test tubes are kept at room temperature for 5 min with occasional shaking.
8. The tubes are placed in a water bath at 80°C for 5 minutes with frequent rigorous shaking.
9. Then take out the test tubes into chilled ice.
10. Add 4 ml of 3N H_2SO_4 and 2 ml of 5 per cent DMEB (4-dimethylamin obenzaldehyde) in n-propanol to each tube.
11. After through mixing, the tubes are again placed in water bath at 70°C for 16 min.
12. Take tubes out and the absorbance is measured at a wavelength of 540 nm in a spectrophotometer.
13. HP content is expressed in mg/100g of tissue, by referring to a standard graph.

For Preparing Standard Graph

1. 50 mg of 4-hydroxypyrrolidine-alpha-carbonic acid (hydroxyproline) was dissolved in 100 ml distilled water by adding 1 drop of H_2SO_4.
2. 5 ml of the above solution is diluted to 50 ml with distilled water to get working standard.
3. Then an actual working standard solution is prepared from this by diluting 10, 20, 30 and 40 ml of this working standard solution to 100 ml with distilled water.
4. From the above diluted working standard a concentration containing 10 µg/ml is selected from which 0.1, 0.2 ml, 0.3ml and 0.4 ml are taken as standards for 1µg, 2µg, 3µg and 4µg hydroxyproline standards.
5. The volume is made up to 1 ml with the distilled water during the experiment.
6. A standard graph is plotted with concentrations of HP against the corresponding absorbance values.

A conversion factor of 7.25 (Goll at al., 1963) is used to estimate the collagen content. The calculation for estimating the hydroxyproline in the meat sample is outlined by Woessner (1961).

$$HP\ (g/100g) = \frac{\text{O.D. unknown} \times \text{Conc. of standard} \times \text{Total volume from which aliquot was taken} \times \text{Total volume made} \times 100}{\text{O.D. standard} \times \text{Aliquot taken} \times \text{Volume of solution used for Neutralizing} \times \text{Weight of the sample taken} \times 1000 \times 1000}$$

$$\text{Collagen content } (g/100g) = HP\ (g/100g) \times 7.25$$

Method–V: Estimation of Collagen solubility

Collagen solubility is estimated by measuring the soluble hydroxyproline fraction in the cooked meat sample. Soluble fraction of hydroxyproline is prepared by slightly modifying the procedure of Okonkwo *et al.* (1992).

Reagents and Materials

1. Ringer solution; (8.6 g NaCl, 0.3 g KCl and 0.33 g $CaCl_2$ per litre)
2. Standard of hydroxyl proline
3. 6N HCl
4. Hot air oven
5. Hot water bath
6. Centrifuge machine
7. Weighing machine
8. Glass wares

Procedure

- ☆ Take 4 g of minced meat samples and place in 50 ml glass centrifuge tubes.
- ☆ Add 12 ml of one-fourth strength Ringer's solution (8.6 g NaCl, 0.3 g KCl and 0.33 g $CaCl_2$ per litre) (Hill, 1966).
- ☆ Heat the samples at 77°C for 70 min in a pre-standardized water bath.
- ☆ Stir the samples occasionally.
- ☆ Centrifuge it at 4000 rpm for 20 min.
- ☆ Then transfer the samples quantitatively into a 50 ml poly-carbonate centrifuge tube.

- ✰ The supernatant is decanted into a 50 ml beaker.
- ✰ Wash the residues with 8 ml one fourth ringer's solution and centrifuge again at 4000 rpm for 10 minutes.
- ✰ The supernatant are combined after centrifugation representing the soluble fraction.
- ✰ The residues remained after centrifugation represented insoluble fraction.
- ✰ Each fraction is separately transferred into a narrow mouthed 50 ml conical/ether extraction flask and covered with watch glass/small petridish.
- ✰ The supernatants and residues are hydrolyzed with 40 ml of 6N HCl for 12 hrs at 105°C in a hot air oven.
- ✰ Hydroxyproline in the hydrolysates were determined by the procedure of Neuman and Logan (1950) as mentioned in the estimation of collagen content.

The percent soluble hydroxyproline is calculated as follows.

$$\text{Per cent soluble hydroxyproline} = \frac{\text{Soluble hydroxyproline of supernatant fraction} \times 100}{\text{Soluble hydroxy proline of supernatant} + \text{Insoluble hydroxy proline of residue fraction}}$$

$$\text{Collagen solubility (\%)} = \frac{\text{Soluble collagen} \times 100}{\text{Soluble collagen} + \text{Insoluble collagen}}$$

32. Estimation of Cholesterol Content

Total cholesterol in Meat products can be determined by using the method of Zlatkis *et al.* (1953) with little modifications as described by Rajkumar *et al.* (2004).

Reagents and Materials

1. Chloroform
2. Methanol
3. Blender
4. Whatman Filter paper no. 42

5. Centrifuge machine
6. Separatory funnels
7. Spectrophotometer

Procedure

- ☆ First prepare the lipid extract. It was prepared by mixing one gram of minced meat with 10 ml of freshly prepared 2:1 Chloroform: Methanol solution.
- ☆ Homogenize it in a blender (Remi Motors Model RQ 127).
- ☆ Filter the homogenate using Whatman filter paper No. 42.
- ☆ Take 5 ml of filtrate and add equal quantity of distilled water, mix it well and centrifuged at 3000 rpm for 7 min.
- ☆ Top layer (methanol) was removed by suction.
- ☆ Volume of bottom layer (Chloroform) having cholesterol was recorded.
- ☆ The O.D. of standard and test against blank was taken at 560 nm.

Total cholesterol mg percent was recorded as follows:

$$\frac{\text{O.D. of sample}}{\text{O.D. of standard}} \times \frac{\text{Vol. of chloroform}}{\text{Weight of sample taken}} \times \frac{1}{\text{Conc. of standard cholesterol in mg \%}}$$

33. Protein Digestibility (*in vitro*)

The *in vitro* digestability of protein is estimated by the method of Akeson and Stahmann (1964) as modified by Singh and Jambunathan (1981).

Reagents and Materials

1. *Pepsin*: 0.4 mg/ml; Forty mg pepsin (3000 units) was dissolved in HCl (0.1N) pH 2.0
2. *Pancreatin*: Fifty mg pancreatin is dissolved in 100 ml of 0.1 M borate buffer (pH 6.8) containing calcium chloride.
3. 10 per cent Trichloroacetic acid (TCA)
4. 5 per cent TCA
5. Sodium Hydroxide (0.2N)

6. 0.025 M Calcium chloride : Calcium chloride (2.75g fused) is dissolved in one litre borate buffer.
7. 0.1M borate buffer (pH 6.8)
   (*a*) *0.2 M boric acid*: Boric acid (12.4g) is dissolved in distilled water and volume is made upto one litre with distilled water.
   (*b*) *0.05M borax solution*: Borax (19.05 g) was dissolved in water and volume is made up to one liter with water. Borate buffer (pH 6.8) containing 0.025M calcium chloride is prepared by adding 140 ml 0.2 M boric acid, 50 ml distilled water and 3 drops of 0.05 M borax solution to set up at 6.8.

0.1M borate buffer (pH 6.8) containing 0.02 M calcium chloride is prepared by dissolving 2.75g calcium chloride in one litre of the buffer.

Procedure

- ✰ Two hundred mg of ground and moisture free sample is taken in a 25 ml conical flask.
- ✰ To this 5 ml pepsin solution is added and incubated at 37°C for 16 h in water bath-cum-shaker.
- ✰ A few drops of toluene are added to each flask to check the growth of micro-organisms.
- ✰ After the pH of the contents is adjusted to 7.0 with 0.2N sodium hydroxide solution, two ml pancreatin solution is added to each conical flask and incubated at 37°C in water bath-cum-shaker for 24 h.
- ✰ Then 7 ml of 10 per cent trichloroacetic acid is added and the content is centrifuged at 12,000 rpm for 20 min and the residue is washed thrice with 5 ml of 5 per cent trichloroacetic acid.
- ✰ The supernatants are pooled and volume is made up to 25 ml with 5 per cent trichloroacetic acid.
- ✰ Fifteen ml aliquot is taken for nitrogen estimation (AOAC, 1980). The protein digestibility of the sample is detrmined by the following formula.

$$\text{Protein digestibility (per cent)} = \frac{\text{Digested Protein}}{\text{Total Protein}} \times 100$$

34. Protein Efficiency Ratio

Method–I

PER of the meat product can be calculated by using varying amounts of collagen in the meat product and computing the following linear regression equation to estimate the PER of meat products from collagen content.

$$\text{PER} = -0.0229\ (\text{collagen per cent}) - 3.1528$$

Computed Protein Efficiency ratio (C-PER) can be derived from collagen content by using the following equation.

$$Y = -0.0229\ X + 3.1528$$

where,

Y= estimated C-PER and X= collagen content (per cent total protein)

Method–II: Protein Efficiency Ratio

PER is determined by the method of Chapmen *et al.* (1959).

Rat feeding trial : The weighed diet is daily given and the unconsumed diet is collected, dried and weighed. Food and distilled water are given *ad libitum*. The rats are initially weighed and then twice a week. The rats are finally weighed after four weeks and the gain in weight of rats during this period was recorded. The amount of food and protein intake during this period are calculated on dry matter basis. Protein efficiency ratio are worked out by the following formulae:

$$\text{PER} = \frac{\text{Gain in body weight (g)}}{\text{Protein consumed (g)}}$$

$$\text{FER} = \frac{\text{Gain in body weight (g)}}{\text{Food consumed (g)}}$$

35. Apparent Digestibility (AD), True Digestibility (TD), Net Protein Utilization (NPU) and Biological Value (BV)

These parameters are determined by nitrogen balanced study using the method of Chick *et al.* (1935). After 28 days of feeding PER, the rats are transferred singly to metabolic cages for the determination of AD, TD and BV and continued to receive same diets.

Another group of rats of same weight and age is fed on a nitrogen free diet. The experiment is conducted for nne days, for first four days rats are acclimatized and for remaining five days urine and faeces are separately collected from each rat. A few drops of H_2SO_4 (10 per cent) were added to urine so as to prevent any loss of ammonia. The faeces and urine of each rat was pooled separately. Samples of faeces are dried at 70°C, weighed and powdered. The urine samples are filtered and transferred to volumetric flasks and volumes were made to 100 ml. Nitrogen in each sample of urine and faeces are determined by Micro-Kjeldahl method (AOAC, 1980). From the data of nitrogen intake and excretion in faeces and urine, apparent digestibility (AD), true digestibility (TD), biological value (BV) and net protein utilization (NPU) are determined. In calculating apparent digestibility, the metabolic faecal nitrogen is not taken into consideration. Apparent digestibility is worked out with the help of following formula:

$$AD = \frac{Ni - Fn}{Ni} \times 100$$

where,

Ni = Nitrogen intake

Fn= Faecal nitrogen

For estimation of true digestibility, the metabolic faecal nitrogen (nitrogen excreted by the nitrogen free group through faeces) is taken into account to find out the exact amount of nitrogen digested. While calculating the BV, the nitrogen excreted by the non-protein group through urine *i.e.* endogenous urinary nitrogen (En) is also considered. TD and BV are worked out using the following formulae:

$$TD = \frac{Ni-(Fn-Mn)}{Ni} \times 100$$

where,

Fn = Faecal nitrogen

Mn = Metabolic faecal nitrogen

$$BV = \frac{Ni-(Fn-Mn)-(Un-En)}{Ni-(Fn-Mn)} \times 100$$

where,

Fn = Faecal nitrogen

Mn = Metabolic faecal nitrogen

Un = Urinary nitrogen

En = Endogenous urinary nitrogen

Net Protein Utilisation (NPU) :

NPU is determined by the method of Platt *et al.* (1961)

$$NPU = \frac{BV \times TD}{100}$$

36. Net Protein Retention (NPR)

The data obtained during the experiment for determining the biological value are used to calculate NPR value. In this case the initial weights of the rats and final weights of rats after the experiment period of five days are recorded. Rats fed on test diets showed increase in the body weights and those fed on nitrogen-free diet showed a decrease in the body weight. The values of NPR are calculated according to Bender and Doell (1957) as follows:

$$NPR = \frac{\text{Gain in wt. of test animals + loss in wt. of control animals}}{\text{Protein consumed by test animals}}$$

37. Protein Retention Efficiency (PRE)

The values of PRE are recorded as follows:

$$PRE = NPR \times 16$$

Utilizable Proteins (UP)

The utilizable proteins are estimated by the method of Gupta *et al*. (1979) as follows:

$$\text{Utilizable Protein} = \frac{\text{NPU} \times \text{n (In Per cent of Dry Matter)} \times 6.25}{100}$$

On the last day of the experiment, rats are weighed and lightly anaesthesized with diethyl ether. Blood samples are withdrawn by decapitation and cardiac puncture with polyethylene syringes. The liver, heart and kidneys are excised, cleared of adhering matter, blotted in filter paper and kept in the deep freeze for further analysis.

38. Hydratability of Meat Products

Hydratability is the important crieteria to determine the texture characteristics of dried meat products. The higher percentage of hydratability lead to lower shelf life of meat and crispiness of the products is lost. Hydratability of the meat sample is determined as weight of water absorbed by the meat (gm)/weight of dry sample. It is generally determined based on the procedure of Mittal and Lawrie (1986).

Reagents and Materials

1. Filter paper
2. Distilled water
3. Weighing scale
4. Hot water bath

Procedure

- ☆ Take about 2.5 gm of carefully weighed meat sample.
- ☆ Place the sample in a test tube with an excess boiling water.
- ☆ Immerse the tube in a boiling water bath for 5 min. to hydrate the meat sample.
- ☆ Drain out the hydrated samples for 5 min., with an intermittent blotting.
- ☆ Carefully weigh the sample again.
- ☆ Hydratability of the meat sample is determined as weight of water absorbed to the weight of the sample and it is expressed as gm/100g of sample weight.

39. Estimation of Water Absorption Index (WAI) of the Extruded Meat Products

Water absorption index is the important factor to determine the processing quality of extruded products. It also help us to assess the shelf life of meat products w.r.t. sensory attributes and subsequently help in the designing of formulations. The water absorption index is determined by the method given by Anderson *et al.* (1969).

Reagents and Materials

1. Centrifuge machine
2. Distilled water
3. Weighing scale
4. Glasswares

Procedure

- ✰ Weigh 2.5 gm of well ground meat sample into 100ml centrifuge tubes.
- ✰ Then add 30 ml of distilled water into it.
- ✰ Keep the sample for 30 min with occasional stirring for equilibration.
- ✰ Then centrifuge it at 5000 rpm for 10 min,
- ✰ Carefully pour the supernatant into petridish and weigh the remaining gel.
- ✰ The water absorption index can be calculated as the ratio of weight of gel obtained to that of initial weight of the sample (g/g).

40. Estimation of Water Solubility Index (WSI) of the Given Meat Sample

The water solubility index was measured according to procedure described by Machado *et al.* (1998).

Reagents and Materials

1. Hot air oven
2. Weighing machine
3. Petri plates

Procedure

1. The supernatant liquid obtained during the determination of Water Absorption Index as described earlier (Anderson *et al.*, 1969).
2. Collect the supernatant liquid in a petri plate.
3. Keep this petriplate into a hot air oven to evaporate it to dryness.
4. After drying, cool the petridishes and weigh them carefully.
5. The water solubility index is determined as weight of solids to the initial weight of the sample (g/g).

41. Water Activity (a_w)

Method–I

Water activity of meat products are measured by a Pawkit water activity meter (Decagon Devices, Pullman, Washington, USA).

Apparatus Required

1. Water activity meter
2. Weighing scale

Procedure

- ☆ Measure the samples in duplicate.
- ☆ The duplicate sample are taken in a sample container provided with water activity meter and placed inside the apparatus.
- ☆ Proper care is taken not to touch the sensor and the upper lid is closed.
- ☆ Then the left button is pressed once to start measuring water activity.
- ☆ The lid is opened and the sample is taken out after getting a beep sound indicating the apparatus had finished estimating a_w.
- ☆ It took 5 minutes to take one reading.
- ☆ The activity meter is calibrated at regular intervals.

Method–II: Measurement of Water Activity (a_w)

Water activity of the cooked meat product is measured indirectly by measuring the equilibrium relative humidity (ERH) with the help of a dry and wet bulb thermometer as described by Labuza *et al.* (1970).

Apparatus Required

1. Relative humidity meter
2. Medium density Polyethylene bags

Procedure

☆ Keep the sample inside a bell jar along with the dry and wet bulb thermometer for 1 h and the readings of the thermometer are recorded.

The water activity value (a_w) is determined by applying standard formula of Landrock and Procter (1951).

$$\text{Water activity } (a_w) = \frac{\text{Perc ent ERH}}{100}$$

where, ERH is the equilibrium relative humidity at which the products gain or loose water.

Method–III: Measurement of Water Activity (a_w)

a_w is determined by using the graphical interpolation method of obtaining equilibrium relative humidity (ERH) described by Landrock and Procter (1951a,b).

Reagents and Materials

1. Saturated salt solutions with different relative humidity as per table
2. Desiccators
3. Porcelain plates with holes
4. Incubation at temperature 25±1°C.

Procedure

☆ Take approximate 50 ml of saturated salt solutions with different relative humidity as per table.

Table 3: Relative Humidity of Air Over Saturated Solution of Salts

Salt	*Relative Humidity per cent at °C*							
	5°	*10°*	*15°*	*20°*	*25°*	*30°*	*35°*	*40°*
Lithium chloride	16	14	13	12	11	11	11	11
Potassium acetate	25	24	24	23	23	23	23	23
Magnesium bromide	32	31	31	31	31	30	30	30
Potassium carbonate	–	47	45	44	43	42	41	40
Magnesium nitrate	54	53	53	52	52	52	51	51
Cupric chloride	65	68	68	68	67	67	67	67
Lithium acetate	72	72	71	70	68	66	65	64
Strontium chloride	77	77	75	73	71	69	68	68
Sodium chloride	76	75	75	75	75	75	75	75
Ammonium sulphate	81	80	79	79	79	79	79	79
Cadmium chloride	83	83	83	82	82	82	79	75
Ammonium sulphate	81	80	79	79	79	79	79	79
Cadmium chloride	83	83	83	82	82	82	79	75
Potassium bromide	–	86	85	84	83	82	81	80
Lithium sulphate	84	84	84	85	85	85	85	81
Potassium chloride	88	87	87	86	86	84	84	83
Potassium chromate	89	89	88	88	87	86	84	82
Sodium benzoate	88	88	88	88	88	88	86	83
Barium chloride	93	93	92	91	90	89	88	87
Potassium nitrate	96	95	97	97	97	97	96	96
Potassium sulfate	98	97	97	97	97	97	96	96
Disodium phosphate	98	98	98	98	97	96	93	91
Lead nitrate	99	99	98	98	97	96	96	95
Zinc sulfate	95	93	92	90	88	87	85	84

Rockland, 1960.

- ☆ Pour these solutions into separate desiccator
- ☆ A porcelain plate with holes was kept over the saturated solution in the desiccated.
- ☆ Weigh exactly 10 g of sample in a dried petridish for each desiccator.

- ☆ Keep this petridish on the porcelain plate.
- ☆ Cover the desiccators with a lid and kept at specified temperature (25±1°C).
- ☆ The samples weighed separately exactly after 6, 12, 24 hours and based on these observations weight gain or loss was calculated and plotted against per cent relative humidity.
- ☆ The point of intersection at the base line of the plots of weight gain or loss gave the corresponding ERH.
- ☆ The water activity can be calculated by using following equation:

$$\text{ERH (per cent)} = a_w \times 100$$

$$a_w = \frac{\text{ERH}}{100}$$

Precautions

1. Take all the desiccators of same size.
2. The petridish should not cover the holes of the porcelain plate.
3. All the porcelain plates should have same size and number of holes.
4. Weigh the sample accurately and immediately after taking out from the desiccators.

Chapter 4
Minerals, Vitamins and Enzymes

1. Minerals Estimation

Sample Solution for the Estimation of Minerals

(*a*) From Dry Ash

Add 20 ml 1:1 hydrochloric acid in the crucible containing white ash. The crucible is covered with a watch glass. The mixture is evaporated to dryness on a boiling water bath. The watch glass is rinsed with a small portion of distilled water and 1 ml conc. HNO_3 is added to the crucible. The solution is evaporated to dryness. Then four ml of 1:1 Hydrochloric acid is added and the solution is warmed over the boiling water bath for a few minutes. It is then filtered into a 100 ml volumetric flask using Whatman's No. 42 filter paper. 2-3 washings are given to the filter paper. After cooling, the volume is made up to 100 ml with distilled water.

(*b*) From Wet Digestion

Weigh 2.5 g sample into 500 ml Kjeldahl flask. Add 20-30 ml HNO3 and boil gently for 30-45 minutes to oxidise all easily oxidisable matter. Cool solution somewhat, add 10 ml 70 per cent perchloric acid. Boil very gently; adjusting flame as necessary, until solution is colorless or nearly so and dense white fumes appear. Use particular care not to boil to dryness at any time. Cool slightly, add

50 ml H_2O and boil to drive out any NO_2 fumes. Cool, dilute and filter into 250 ml volumetric flask through Whatman's filter paper 42. 2-3 washings are given to the filter paper with distilled water. Finally, the volume is made with distilled water.

1. Determination of Iron

Principle

The organic substance present in a given material is destroyed by dry ashing. The iron present in ash is converted from ferrous to ferric form by hydroxylamine hydrochloride and then into colored complex by orthophenanthroline. The color thus, developed is measured at 510 nm. The optical density so obtained is proportional to concentration of iron.

Reagents and Materials

1. Orthophenanthroline solution 0.1 per cent.
2. Hydroxyamine Hydrochloride 10 per cent.
3. Acetate buffer solution:
4. Iron standard solution
5. Spectrophotometer
6. Volumetric flasks

Acetate Buffer Solution

Dissolve 8.3g anhydrous sodium acetate (previously dried at 100°C) in H_2O, add 12 ml acetic acid and dilute to 100 ml.

Iron Standard solution

1. Dissolve 3.512 g Ammonium ferrous sulphate in H_2O, add 2 drops of HCl and dilute to 500 ml (1mg/ml).
2. Working standard: Dilute 10 ml of the above solution to 1 litre (10 µg/ml).

Procedure

Preparation of Sample

- ☆ Take 10, 20, 30, 40, 50 ml of the above working standard into 100 ml volumetric flask.
- ☆ Add 2 ml HCl and make volume with distilled water.

- ☆ For blank take 2 ml HCl in another 100 ml volumetric flask and dilute to the volume with H_2O.
- ☆ Take 10 ml each of standards, blank and sample in duplicate in 25 ml volumetric flasks.
- ☆ Add 1 ml Hydroxylamine HCl to each of flasks. Then immediately, add 5 ml acetate buffer solution.
- ☆ After 10 minutes add 1 ml orthophenanthraline solution to each of flasks and shake well. After 10 minutes dilute to volume of each and shake well again.
- ☆ Read the color intensity at 510 nm in Spectrophotometer.
- ☆ Prepare the standard curve.

Calculations

W = Let weight of sample taken for HCl extract

V = Volume of HCl extract made

10 ml = Aliquot of HCl extract taken

m = Conc. of sample aliquot obtained from the standard curve

$$\text{Then Iron (mg/100 g)} = \frac{m \times V \times 100}{W \times 1000} \times 100$$

Preparation of Sample

Take 2.5 g of well dried sample in a crucible and ignite to a clean and white ash in a muffle furnace at 550°C. Cool and prepare HCl extract as described earlier.

Note

- ☆ Make all glasswares *i.e.*, volumetric flasks, pipettes, beakers etc. iron free by dipping into dilute HCl overnight.
- ☆ Filter all reagents to remove suspended matter.

2. Determination of Phosphorus

Method–I

Principle

Orthophosphate in the ash react with Ammonium molybdate in acid solution to form phosphomolybdic acid solution. This compound is reduced by amino-naphtholsulphonic acid reagent to give an intense blue color, which is measured colorimetrically.

Reagents and Materials

1. Perchloric acid 70 per cent.
2. Ammonium Molybdate 5 per cent.
3. *1:2:4 Amino-naphthol sulphonic acid reagent*: Dissolve 30 g sodium disulphite, 6g sodium sulphite and 0.5 g 1:2:4 Amino-naphthol sulphonic acid in distilled water and dilute to 250 ml. Filter the reagent if there remains suspension.
4. *Standard phosphorus*: Dry potassium dihydrogen orthophosphate for 2 hours at 105°C. Weigh 2.1956 g of KH2PO4 and dissolve in water and dilute to 500 ml. (1mg/ml). Dilute 10 ml of the phosphorus stock solution to 1 litre with water (10µg/ml).

Procedure

Preparation of Sample Solution

- ☆ Weigh 2.5 g of well dried and ground sample in a crucible.
- ☆ Ash in a muffle furnace at 550°C.
- ☆ Prepare HCl extract as described earlier.

Development of Color

- ☆ Pipette 0.4, 0.8, 1.2, 1.6 and 2.0 ml of working standard solution (10 µg/ml) in 25 ml volumetric flasks in duplicate.
- ☆ Also take 0.2 ml HCl extract of the sample and 1 ml distilled water for blank in 25 ml volumetric flasks.
- ☆ To each of flasks add 1 ml of 70 per cent perchloric acid, shake and add 1 ml Ammonium molybdate solution.
- ☆ Again shake and after 10 minutes add 0.5 ml Amino-naphthol sulphonic acid reagent. Shake the contents well and make volume upto 25 ml.
- ☆ Read the color intensity at 720 nm.
- ☆ Prepare a standard curve and find the concentration of the sample aliquot.

Calculations

W g = Let weight (g) of sample taken for HCl extract

V ml = Volume of HCl extract made

0.2 ml = Aliquot of HCl extract used for color development

m g = Concentration of sample aliquot obtained from graph

$$\text{Phosphorus (mg/100g)} = \frac{m \times V \times 100}{0.2 \times 1000 \times W}$$

Method–II: Phosphorus Estimation

The phosphorus content was determined by the method of Jackson (1962).

Reagents and Materials

1. *Standard phosphorus solution*: 0.2195 g of Potassium dihydrogen phosphate was dissolved in about 400 ml of distilled water. The 25 ml of 7N H_2SO_4 was added and the solution made to 1000 ml volume and mixed, containing 50 ppm of phosphorus. This solution was diluted directly for the yellow colored solutions of the Vanadomolybdophosphoric acid.
2. *Combined HNO_3–vanadate*: molybdate reagent:

   Solution A: 25 g of Ammonium molybdate was dissolved in 400 ml of water.

   Solution B: 1.25 g of Ammonium metavanadate in 300 ml of boiling water. 250 ml of concentrated HNO_3 was added to the solution and solution B was cooled to room temperature. Finally solution A was poured into B and the mixture was diluted to one litre.

Procedure

Preparation of Test Solution

- An aliquot of phosphorus solution was placed in a 50 ml volumetric flask in a volume not exceeding 35 ml.

Vanadomolybdophosphoric Yellow Color

- To the above phosphorus solution, 10 ml of the vanadomolybdate reagent was added and the solution was diluted to 50 ml with distilled water and mixed well.

- ✰ The developed color was read in the colorimeter by using a filter of 470 nm wavelength against a suitable blank.
- ✰ The standard curve was plotted using known concentration of phosphorus

3. Determination of Calcium

Method–I: Titration Method

Principle

The calcium in the sample solution reacts with Ammonium oxalate to form calcium oxalate at 4.5 pH. The precipitate of calcium oxalate are reduced by $KMnO_4$ solution in acidic medium. The end point being light pink in color. From the volume of $KMnO_4$ used in titration, the concentration of calcium can be calculated.

Reactions

$$CaCl_2 + \begin{array}{c} COONH_4 \\ | \\ COONH_4 \end{array} \longrightarrow \begin{array}{c} COO \\ \\ COO \end{array}\!\!\!> Ca + 2NH_4Cl$$

$$2KMNO_4 + 3H_2SO_4 \longrightarrow K_2SO_4 + 2MnSO_4 + 3\,H_2O + 5\,\{O\}$$

$$\begin{array}{c} COO \\ \\ COO \end{array}\!\!\!> Ca + H_2SO_4 \longrightarrow CaSO_4 + \begin{array}{c} COOH \\ | \\ COOH \end{array}$$

$$\begin{array}{c} COOH \\ | \\ COOH \end{array} + [0] \longrightarrow 2CO_2 + H_2O$$

Reagents and Materials

1. Saturated Ammonium oxalate solution (25 per cent).
2. Ammonia solution 1:4 dilution.
3. N/10 Sodium oxalate: Dissolve 1.675 g of sodium oxalate in H_2O to make volume 500 ml.
4. N/20 $KMnO_4$: Dissolve 0.80 g of $KMNO_4$ in H_2O. Heat to boil for a few minutes. Dilute to 500 ml. Standardize $KMnO_4$ solution against N/20 sodium oxalate solution.

Procedure

Preparation of Sample Solution

- ☆ Weigh 5g of well dried sample into a crucible.
- ☆ Ignite the sample to a white clean ash and prepare the HCl extract as described earlier.

Precipitation of Calcium

- ☆ Take 25 ml of the HCl extract into a 400 ml beaker and add 50 ml H_2O.
- ☆ Add 8-10 drops of methyl red indicator and 10 ml of saturated ammonium oxalate.
- ☆ Cover the beaker with watch glass and heat to boiling point.
- ☆ Precipitate calcium by adding 1:4 Ammonia solution drop-wise until pH is 4.4–4.6 as indicated by pink color.
- ☆ Boil 1-2 minutes and let precipitate settle until clear or overnight.
- ☆ Filter supernatant through Whatman filter paper No. 42.
- ☆ Wash beaker and pipette with 50 ml dilute ammonia solution in small portions.
- ☆ Then wash with hot water until become chloride free, which can be tested by filtrate dilute $AgNO_3$ solution.
- ☆ Break point of filter paper and wash filter paper with 100 ml of water and 5 ml of H_2SO_4 at 80-90°C. Transfer the filter paper into the beaker also.
- ☆ Titrate the contents at 60°C to 70°C with N/20 $KMnO_4$ until slight pink color is obtained.
- ☆ For blank titrate 100 ml H_2O and 5 ml H_2SO_4 against N/20 $KMnO_4$.

Calculations

W = Let weight (g) of sample taken

V_1 = Volume (ml) of HCl extract made

V_2 = Aliquot of HCl Extract used for precipitation

V_3 = Volume of N/20 $KMnO_4$ used against sample aliquot

V_4 = Volume of N/20 $KMnO_4$ used against blank

As 1 ml of N/20 $KMnO_4$ contains 1 mg of Ca, Then

$$\text{Calcium (mg/100g)} = \frac{(V_3 - V_4) \times 1 \times V_1 \times 100}{W \times V_2}$$

Method–II: Minerals Estimation

Digestion

- ✰ Take One gram grounded sample in a conical flask. To this add 25-30 ml diacid mixture HNO_3:$HClO_4$::5:1 (v/v) and kept overnight.
- ✰ It was digested next day by heating till clear white precipitates settled down at bottom of conical flask.
- ✰ The crystals were dissolved by diluting in double distilled water.
- ✰ The contents were filtered through Whatman filter paper No. 42.
- ✰ The filtrate was made to 50 ml with double distilled water and used for determination of total calcium and phosphorus.

Calcium Estimation

The calcium content was determined by the method of Cheng and Bray (1951).

Reagents and Materials

1. *Ammonium chloride*: Ammonium hydroxide buffer solution: 67.5 g of ammonium chloride was dissolved in 570 ml of concentrated ammonium hydroxide and made to 1 litre.
2. *Sodium hydroxide (4N)*: 160 g of sodium hydroxide was dissolved in 1 litre of water.
3. *Standard calcium chloride solution (0.01N)*: 0.5 g of pure Calcium carbonate in 10 ml of 3N Hydrochloric acid was dissolved and diluted to 1 litre volume.
4. *Erichrome black T indicator*: 0.5 g of Erichrome black T and 4.5 g of hydroxylamine hydrochloride were dissolved in 100 ml of 95 per cent ethanol.

5. *Ammonium purpurate indicator*: 0.5 g of ammonium purpurate was mixed with 100 g of powdered potassium suphate.
6. *Ethylenediaminetetraacetate solution (0.01N EDTA)*: 2.00 g of disodium dihydrogen ethylene diamine tetra acetate and 0.05 g of magnesium chloride hexahydrate were dissolved in water and diluted to 1 litre.

Procedure

- ✰ Pippetted 5 ml aliquot into a porcelain dish.
- ✰ Diluted to a volume of approximately 25 ml.
- ✰ Added 0.25 ml of Sodium hydroxide (4 N) and approximately 50 mg of Ammonium purpurate indicator.
- ✰ Titrated with Ethylene diamine tetra acetate (0.01N EDTA) solution, using a 10 ml microburette.
- ✰ The color change was orange to lavender or purple. When closed to the end point, EDTA was added at the rate of about a drop every 5 to 10 seconds, as the color change was not instantaneous.
- ✰ A blank containing Sodium hydroxide, Ammonium purpurate indicator and a drop or two of EDTA aided in distinguishing the end point.

Calculations

$$\text{Calcium (per cent)} = \text{ml EDTA} \times \frac{0.0481}{\text{Wt. of sample (g)}}$$

Method–III: Estimate the Calcium Content by Flame Photometry Standard Curve

Principle

The calcium salt in the solution of HCl extracted is atomized by the burner of the flame photometry, resulting the excitation of the atoms which exit radiant energy of specific wave length of color as determined by the detector, and the reading is taken by the reader of flame photometer, then interpreted the standard curve obtained before, to know the calcium content of the given food sample.

Reagents and Materials

1. Flame photometer
2. Fuel gas.
3. Hydrochloric acid extract of the given food sample

Procedure

- ☆ The estimation of calcium content was determined taking directly the HCl extract of the food sample.
- ☆ To start with first the fuel gas was made on.
- ☆ Then burner of the flame photometer was lighted.
- ☆ The pressure of the gas was correctly adjusted to get the desired flame of light.
- ☆ Then the switch of the compressor was made on.
- ☆ Specific filter for Ca estimation was set in. First of all to set zero, simply distilled water was taken and the readings was adjusted to zero.
- ☆ Then the maximum concentration of the standard solutions already prepared was taken and with it, properly adjusted for sensitivity of Flame Photometer to set 100.
- ☆ Similar samples of with minimum and maximum concentration of calcium was taken and adjusted and observed for 2 or 3 times repeatedly to know whether flame photometer is sensitive to set zero and to set 100.
- ☆ After such adjustment first the readings for the standard curve was noted starting from the lowest concentration (2mg/ml) to highest concentration (26 mg/ml).
- ☆ Then graph was plotted for the standard curve taking concentration of the solution (mg/ml) in X-axis and Flame Photometer reading in Y-axis.
- ☆ Then the HCl extract was taken which contained the dissolved calcium salt of the sample.
- ☆ This solution was spread into finer partially over the flame. The calcium atom was excited, then electronic imbalance and the atoms emitted radiant energy of specific wave length for calcium which was sensitized and detected by Flame Photometer reader.

- ☆ This reading was noted and recorded on the graph of the standard curve for which the calcium content was easily calculated.

Reactions

$$\begin{array}{l} COONH_4 \\ | \\ COONH_4 \end{array} + CaCl_2 \longrightarrow \begin{array}{l} COO \\ COO \end{array}\!\!\!> Ca + 2\,NH_4Cl$$

$$2KMnO_4 + 3H_2SO_4 \longrightarrow K_2SO_4 + 2MnSO_4 + 3H_2O + 5(O)$$

$$\begin{array}{l} COO \\ COO \end{array}\!\!\!> Ca + H_2SO_4 \longrightarrow CaSO_4 + \begin{array}{l} COOH \\ | \\ COOH \end{array}$$

$$\begin{array}{l} COOH \\ | \\ COOH \end{array} + [O] \longrightarrow 2CO_2 + H_2O$$

Procedure

To Prepare Standard Curve

- ☆ 5, 10, 15, 20, 25, 30, 35, 40, 45 and 50 ml of working stock solution (100 ml dil)–10 ml of it were transferred to 100 ml volumetric flask.
- ☆ Then 2 ml of concentrated HCl was added to each flask and the volume was made upto 100 ml.

Color Development

- ☆ 10 ml of the HCl extract of the food sample was taken into a 25 ml volumetric flask.
- ☆ Similarly for standard curve also, 10 ml of the standard solution from each of the flask prepared earlier was taken.
- ☆ For blank also 10 ml dilute HCl was taken (2 ml concentrated HCl in 100 ml distilled water)
- ☆ To all these flasks 1 ml of Hydroxylamine HCl solution was added, waited 5 min to convert the iron present to ferrous form

- ☆ Then 5 ml of buffer solution and 1 ml of 0.1 per cent Orthophenanthroline solution was added.
- ☆ The volume was made to 25 ml with distilled water.
- ☆ After each addition the solution was shaken.
- ☆ Then the color intensity was measured after 5 min at 510 nm.

4. Determination of Copper in the food Sample

Principle

Dithizone is the reagent, which is used in determination for extraction of the metals. The metals are soluble in Dithizone. The sample extracted with 1 per cent dithizone contains metals namely Cu, Zn, Pb. In acidic conditions, Copper goes along with dithizone dissolved in chloroform. Zn, Pb remains in aqueous layer. From residue, after evaporation of CH_3Cl, copper estimation is done by adding Sodium diethyl dithiocarbamate, which imparts yellow to brown color.

Procedure

Purification of Dithizone

- ☆ In 1 gm dithizone, 75 ml $CHCl_3$ was added
- ☆ Extraction was done with 100 ml NH_3 solution (1 per cent) for 4 times
- ☆ Impurity will be removed and only dithizone will go to ammonia layer.
- ☆ Chloroform layer is discarded and alternative layer is extracted 4 times with 1 per cent Ammonia solution.
- ☆ 25 ml of 1 per cent HCl is added and is reextracted with chloroform (4 times).
- ☆ It is washed with water and $CHCl_3$ is evaporated from the 1 per cent solution of dithizone prepared in chloroform
- ☆ 10 ml sample is taken and 2 g citric acid is added. It is neutralized by ammonia.
- ☆ 0.5 ml NH_3 is added in excess
- ☆ It is extracted thrice with 10, 5 and 5 ml of 0.1 per cent solution of dithizone

- ✰ It is washed with water.
- ✰ In dithizone layer, 2 ml of 1 per cent HCl + 20 ml H_2O is added.
- ✰ It was extracted twice.
- ✰ Aqueous layer contains Zn + Pb.
- ✰ The other layer containing copper.
- ✰ CH_3Cl is evaporated.
- ✰ Volume is made 50 ml with 1 per cent HCl.
- ✰ The Blank is run in the similar manner.
- ✰ Three tubes containing sample (5 ml), Blank, standard (1ml) are taken.
- ✰ In this, 1 ml of 5 per cent citric acid is added and is neutralized with ammonia.
- ✰ 0.5 ml NH_3 is added in excess.
- ✰ 5 ml of gum solution and 5 ml of 0.2 per cent Sodium diethyl dithiocarbamate was added in each tube.
- ✰ The color development was yellow to brown whose intensity was read in colorimeter.

Calculations

Amount of copper per 100 ml of sample =

$$100 \times \text{optical density} \times \text{concentration}$$

5. Determination of Tin in Canned Products

Principle

Organic matter present in product is destroyed by digestion with concentrated H_2SO_4 and concentrated HNO_3 or by ashing. H_2O_2 is used for complete oxidation. Tin is reduced to stannous form by nascent hydrogen produced in reaction. It is titrated with Potassium iodate in the presence of KIO_3 using starch as the indicator. Stannous is converted to stannic form by Iodine. Excess of iodine gives blue color with starch indicator.

Reagents and Materials

1. HCl (1:3)
2. H_2SO_4 (Concentrated)

3. Concentrated HNO_3
4. H_2O_2 (30 per cent)
5. Aluminum foils (A.R. quality)
6. 5 per cent Sodium Carbonate
7. Antimony Chloride, 1.5 gm substance in 100 ml 1:3 HCl.
8. Potassium Iodide
9. Potassium Iodate-N/200
10. Starch Indicator

Starch Indicator

Potassium Iodide solution is prepared by dissolving 0.2g KI and 3g $NaHCO_3$ in 100 ml of boiled and cooled water. 1:3 HCl is added. Potassium Iodate (N/10) is prepared by adding 0.3566 g KIO_3 in 100 ml water. 5 ml is taken from this and made volume to 100 ml (N/200)

Procedure

- ☆ Extract is prepared by digestion or by ashing.
- ☆ Digestion is done as follows: 20 ml concentrated HNO_3 add 10 ml of concentrated H_2SO_4 is added to the food sample and kept in Kjeldahl flask.
- ☆ It is heated till white fumes appear.
- ☆ Add 5 ml of nitric acid drop wise.
- ☆ Add drop wise 10 ml of H_2O_2.
- ☆ It is heated again till white fumes appear
- ☆ The colorless digest is made 100 ml in volume.

Ashing

- ☆ 20 g of sample is taken in silica dish
- ☆ 2g each of Potassium hydrogen phosphate and Magnesium nitrate is added and mixed with slight distilled water.
- ☆ Charred products is kept in muffle furnace at 550° C for 3-4 h and thereafter cooled.
- ☆ Few drops of concentrated HNO_3 is added and is evaporated likewise.

- 5-10 ml H_2O_2 is added.
- It is heated to remove the nitrate fumes.
- 0.5 ml of concentrated HCl is added and evaporated.
- 5ml of conc. HCl is added made upto known volume.
- 25 ml of digested solution is transferred to reduction flask.
- 30ml of 1:3 HCl was added.
- Further 2 drop of $SbCl_3$ and 0.3g of Aluminium foil was added.
- A delivery tube was fitted in the flask.
- The other end of delivery tube was dipped in $NaHCO_3$ solution.
- Initially it is gently heated.
- Later, it is heated till perfectly clear solution is obtained.
- It was cooled in ice water.
- 4 ml of Iodide carbonate reagent was added.
- Starch was added (few drops).
- It was titrated against N/200 KIO_3 solution.

Preparation of Standard Solution

Stock Solution

- 1mg tin/ml dissolved
- 0.19 g of stannous chloride in 5ml concentrated HCl and made upto 100 ml with water.
- 10 ml of stock solution is taken and volume is made 100 ml.
- It gives 1ml =0. 1mg.

Observations

A = Weight of Sample

B = Volume of solution taken for reduction

C = Volume make up

= Titre value

Calculations

$$\text{Tin content (mg/100gm)} = \frac{0.297 \times \text{Titre value} \times \text{Volume made up} \times 100}{\text{Wt. of sample} \times \text{Volume of ash solution taken for reduction}}$$

6. Metals (Atomic Absorption Method)

Application

The determination of calcium, copper, iron, magnesium, manganese, potassium, sodium and zinc in food products.

Principle

After removal of organic matter by dry ashing or wet digestion, the residue is dissolved in dilute acid. The solution is sprayed into the flame of an atomic absorption or emission of metal to be analysed is determined at a specific wavelength.

Reagents and Materials

1. Hydrochloric acid, 6N, 3 N and 0.3 N.
2. Lanthnium chloride 10 per cent W/V.
3. Deionized distilled water.
4. Filter paper Whatman No. 41. Wash the filter paper with 3 N HCl to remove traces of metals.
5. Standard solutions
   (*i*) *Calcium solutions*: (a) Stock solution 25 µg/ml:
   (*ii*) Dissolve 1.249 $CaCO_3$ in 3 N HCl. Dilute to 1 litre. Again dilute 50 ml to 1 litre.
   (*iii*) *Working standard solutions*: 0.5, 10, 5 and 20 µg calcium/ml. Take 5 ml La stock solution into 25 ml volumetric flasks and dilute to 25 ml.
   (*iv*) *Copper stock solution*: (1000 µg/ml). Dissolve 1 g pure copper metal in minimum amount of HNO_3 and add 5 ml HCl. Evaporate almost to dryness and dilute to 1 litre with 0.1 N HCl.
   (*v*) *Iron stock solution*: (1000 µg/ml). Dissolve 1 g pure Fe wire in 30 ml 6 N HCl with boiling. Dilute to 1 litre.

(*vi*) *Magnesium stock solution*: (1000 μg/ml). Place 1,000 g pure Mg metal in 50 ml water and slowly add 10 ml HCl. Dilute to 1 litre.

(*vii*) *Manganese stock solution*: (1000 μg/ml). Dissolve 1.582 g MnO2 in 30 ml 6 N HCl. Boil to remove Cl and dilute to 1 litre.

(*viii*) *Zinc stock*: (1000 μg/ml) : Dissolve 1,000 g pure zinc metal in 10 ml 6 N HCl. Dilute to 1 litre.

Operating Parameter

Element	*Wavelength A*	*Flame*	*Range μg/ml*	*Remarks*
Ca	4227	Rich Air C_2H_2	2-20	1 per cent a, 1 per cent HCl require special
	4227	Rich N_2O-C_2H_2	2-20	
Cu	3247	Air-C_2H_2	2-20	
Fe	2483	Rich Air C_2H_2	2-20	
Mg	2852	Rich Air C_2H_2	2-20	
Mn	2795	Air-C_2H_2	2-20	May need La
Zn	2138	Air-C_2H_2	0.5-5	

Procedure

(a) Solution Obtained from Wet Digestion

- ☆ Transfer the contents of digestion flask (wet digestion) into a graduated flask and dilute to wash with water.
- ☆ Mix thoroughly.

(b) Ash Obtained from Ashing

- ☆ Treat the ash with 1 + 1 HCl (6N) to wet it completely and carefully take to dryness on a steam bath.
- ☆ Add 15 ml of 3 N HCl and heat to just boil. Cool and filter through Whatman No. 41 retaining as much solids in the dish as possible.
- ☆ Add 10 ml of 3 N HCl to the dish and heat just to boil. Cool and filter into a volumetric flask.
- ☆ Wash the dish three times with water.

- ☆ If calcium is to be determined add 5 ml $LaCl_3$ solution per 100 ml of the solution.
- ☆ Cool and dilute to a definite volume with water.
- ☆ Prepare a blank similarly as 1 to 6 except taking ash.

(c) Calibration of Apparatus

- ☆ Set the apparatus according to the instructions.
- ☆ Measure the calibration solutions and reagent blank solution.
- ☆ While running the samples, periodically check that the calibration values remain constant. Flush burner with water between samples.
- ☆ Prepare a calibration curve by plotting the absorption or emission values against standard concentration in μg/ml.

Calculations

Read from the graph the metal concentration (μg/ml) that correspond to the absorption or emission values of the sample and blank.

W = Let weight (g) of sample
V = Volume (ml) of extract
a = Concentration (μg/ml) of sample solution from graph
b = Concentration (μg/ml) of blank solution from graph

$$\text{Then Metal content } (\mu g/ml) = \frac{a-b}{10-W} = V$$

7. Determination of Chloride/Sodium Chloride Content (ISI Handbook Sp : 18 Part XII-1984)

Principle

Extraction of meat or meat product with hot water and precipitation of proteins. After filtration and acidification, addition of an excess of silver nitrate solution to the extract and titration with potassium thiocyanate solution.

Reagents and Materials

1. Nitrobenzene.
2. Nitric acid solution–4 N; mix 1 volume of concentrated HNO_3 (P_{20} = 1.39 to 1.42 gm/ml) with 3 volumes of water.
3. Solutions used for Precipitation of proteins
   (a) Reagent I: Dissolve 106 g of potassium ferrocyanide (II) trihydrate ($K_4 Fe (CN)_6 . 3 H_2O$) in water and dilute to 1000 ml.
   (b) Reagent II: Dissolve 220g of zinc acetate [Zn $(CH_3COO)_2$. $2H_2O$] and 30 ml of glacial acetic acid in water and dilute to 1000 ml.
4. Standard silver nitrate solution–0.1N; dry the silver nitrate ($AgNo_3$) for hours at 150°C and allow to cool in a desiccator. Dissolve 16.989 g of the dried salt in water and dilute to 1000 ml.
5. Standard Potassium thiocyanate solution–0.1N; Dissolve about 9.7 g of potassium thiocyanate (KCNS) in water and dilute to 1000 ml. Standardize the solution to the nearest 0.0001 N against silver nitrate using the solution standard silver nitrate as in '4' and the indicate solution–as specified in '7' *i.e.* Ammonium Ferric (III) sulphate [NH_4 $Fe(SO_4)_2$. $12H_2O$] solution saturated in water.
6. Sodium Hydroxide–1N solution
7. Ammonium Ferric (III) sulphate ($NH_4Fe(SO_4)_2$. $12H_2O$] solution saturated with water.
8. Activated charcoalMeat miner–plate size 4 mm.
9. Volumetric flask–200 ml, one mark.
10. Conical flask–250 ml.
11. Burette–25 or 50 ml
12. Pipettes–20 ml, one mark.
13. pH meter.

Procedure

- ☆ Preparation of sample: As described earlier.
- ☆ Test portion: Weigh, to the nearest 0.001g about 10g of the prepared sample and transfer it quantitatively in a conical flask.

Deproteination

- ☆ Add successively 0.5 g of activated charcoal and 100 ml of hot water to the test protein in the flask.
- ☆ Heat the flask with contents for 15 minutes in a boiling water bath.
- ☆ Shake the contents of the flask repeatedly.
- ☆ Allow the flask with contents to cool to room temperature then add successively 2 ml of Reagent I and 2 ml of Reagent II.
- ☆ Mix thoroughly after each addition.
- ☆ By using sodium hydroxide solution adjust the pH between 7.5 and 8.3 using the pH meter.
- ☆ Allow the flask to stand for 30 minutes at room temperature, transfer the contents quantitatively to the volumetric flask and dilute to the mark with water.
- ☆ Mix the contents thoroughly and filter through a fluted filter paper.

Note

- ☆ The filtrate may also be used for the determination of the nitrate and nitrite content.
- ☆ If ascorbic acid is present in quantities below 0.1 per cent in the sample or if the extract is to be used only for the chloride determination.
- ☆ The activated charcoal can be omitted in preparing the extract.
- ☆ Furthermore, the pH adjustment is not necessary if only a chloride determination is to be performed.

Determination

- ☆ Transfer 20 ml of the filtrate to a conical flask by means of a pipette and add 5 ml of nitric acid solution and 1 ml of indicator solution '7' using a measuring cylinder.
- ☆ Transfer 20 ml of silver nitrate solution to the conical flask by means of a pipette.
- ☆ Add 3 ml of nitrobenzene by means of a measuring cylinder and mix thoroughly.

- ✰ Shake vigorously to coagulate the precipitate.
- ✰ Titrate the content of the conical flask with potassium thiocyanate solution.
- ✰ Record to the nearest 0.02 ml the volume of the potassium thiocyanate solution required.
- ✰ Carry out two determinations on the same prepared sample.

Observations

V = Volume in ml, of 0.1 N potassium thiocyanate solution required for the titration and

M = Mass in g, of the test portion.

Calculations

Chloride content of the sample, expressed as a percentage of sodium chloride by mass, is equal to:

$$0.005844\,(20-V) \times \frac{200}{20} \times \frac{100}{M}$$

$$= 58.44 \times \frac{2-0.1V}{M}$$

Note

- ✰ If the standard volumetric solution of potassium thiocyanate solution is not exactly 0.1000 N, a suitable connection factor should be used in calculating the result.
- ✰ Take as the result the arithmetic means of the two determinations, if the requirements of repeatability as below satisfied. Report the result to the nearest 0.05g per 100 g of sample.
- ✰ *Repeatability*: The difference between the results of two determinations carried out simultaneously or in rapid succession by the same analyst should not be greater than 0.2 g of sodium chloride per 100 g of sample.

8. Bioavailability of Minerals

In vivo Method

One week before the completion of the experiment, rats are placed individually into polyethylene metabolic cages. Faeces and urine are collected separately for each rat during the last 5 days of the experiment. The amount of food consumed during this period was also recorded. The concentrations of calcium, phosphorus, iron, zinc, copper and manganese in food faeces and urine are estimated. From these data, apparent absorption, balance and apparent retention are calculated by following formulae:

$$\text{Apparent absorption (per cent)} = \frac{\text{I–Fi}}{\text{I}} \times 100$$

Balance = I–Fi–Ui

$$\text{Apparent Retention (per cent)} = \frac{\text{I–Fi–Ui}}{\text{I}} \times 100$$

where,

I = Mineral content of food

Fi = Mineral content of faeces

Ui = Mineral content of urine

2. Vitamins Estimation

1. Estimation of β-Carotene Using Column Chromatography

Principle

The individual carotenoid are separated on a column of alumina and determined spectrophotometrically.

Reagents and Materials

1. 3 per cent Acetone in petroleum ether.
2. Alumina (Aluminium oxide neutral).
3. Sodium sulphate anhydrous.

4. β-carotene standard*
5. 0.5 N alcoholic KOH**

* Standard β-carotene solution–Dissolve 50 mg β-carotene in 3 per cent acetone in petroleum ether and dilute to 50 ml volume (1 mg/ml).

** 0.5 N alcoholic KOH: Dissolve 7 g of KOH in ethanol and dilute to 250 ml with ethanol.

Procedure

Preparation of Chromatography Column

- ☆ Fill the chromatography column with aluminium oxide, which has been dried in an oven.
- ☆ Gently press the adsorbent down to a depth of 10 cm by temping device or suction may be used.
- ☆ Cover with a 1 cm layer of anhydrous sodium sulphate.
- ☆ Wet the column with 3 per cent acetone in petroleum ether.
- ☆ Don't let it dry at any stage.

Presentation of Sample

(a) For animal and dairy products (Hot saponification)

- ☆ A known weight of the sample (2-3 g) is taken in a 250 ml round bottom flask and add 15 ml 0.5 N alcoholic KOH for each gram of the sample.
- ☆ Saponify the contents by refluxing on a hot water bath for 30 minutes.
- ☆ Cool and transfer to separatory funnel, adding equal volume of water as wash.
- ☆ Extract four times with 75 ml of freshly redistilled ether extract and discard the aqueous phase.
- ☆ Wash the combined ether extract once more with 50 ml of water so that the aqueous phase gives the red color with phenolphthalein.
- ☆ Dry the ether extract with anhydrous sodium sulphate and evaporate to dryness on a water bath removing the last few millimeters with a stream of nitrogen.

- ☆ Dissolve the residue immediately in 3 per cent acetone in petroleum ether or redistilled chloroform (for Vitamin A) to a definite volume.

(b) Fresh plant materials and food stuffs

- ☆ Finely cut out material with scissors or knife or grind in food chopper.
- ☆ Weigh 10-20 g and add 30 ml acetone-ether (1 per cent).
- ☆ Allow to stand overnight or homogenize in a high speed blender adding 5 g of sodium sulphate.
- ☆ The extract is then filtered and the residue is washed with 3 per cent acetone in petroleum ether until the filtrate is clear of yellow color.
- ☆ The filtrate is pooled and taken in a 500 ml separating funnel.
- ☆ It is then shaked with 50 ml of water.
- ☆ Water washings are discarded.
- ☆ This is repeated 2-3 times in order to make it acetone free.
- ☆ The solvent is then dried over Na_2SO_4 and dilute to 100 ml.

Chromatographic Separation of β-Carotene

- ☆ Take 10 ml of the extract and evaporate it to 2 ml.
- ☆ Pour the condensed extract into the absorption column followed by 10 ml 3 per cent acetone petroleum ether.
- ☆ Collect the elute which contains all the carotene.
- ☆ Transfer the elute to a 25 ml volumetric flask and dilute with acetone petroleum ether (3 per cent).

From the standard curve the concentration of B-carotene in the sample can be calculated.

Calculations

W = Net weight (g) of sample taken

V_1 = Volume of extract made

V_2 = Volume of extract taken for elution

m = Concentration of extract elute obtained from graph

$$\text{Then } \beta\text{–carotene (mg/100g)} = \frac{m \times V_1 \times 100}{V_2 \times W}$$

2.Estimation of Vitamin A (Manual Method)

Principle

The sample is saponified and the unsaponified material is then extracted with diethyl ether. Vitamin A is separated from carotenoids by alumina chromatography. Vitamin A diluted in chloroform antimony trichloride solution is measured spectrophotometrically.

Reagents

1. *0.5 N alcoholic KOH*: Dissolve 7 g of KOH in ethanol and dilute to 250 ml with ethanol.
2. *Chloroform*: Redistilled and dried over Na_2SO_4
3. *Antimony trichloride*: Add 100 g of antimony trichloride to 300 ml of dried chloroform. Heat to dissolve under reflux. Cool and add 7.5 ml of acetic anhydride and dilute to 500 ml with chloroform. Store in a dark bottle.
4. *Vitamin A standard*: Dissolve 25 mg of Vitamin A in chloroform A and dilute to 500 ml with alcohol (500 μg/ml). Take 6 ml of the stock solution of Vitamin A and dilute to 100 ml with chloroform (3 μg/ml).

Procedure

- ✰ Preparation of sample: As described in the hot saponification of sample in β-Carotene.
- ✰ Set the instrument at 100 per cent transmittance with a solution containing 1 ml of chloroform and 9 ml of antimony trichloride reagent.
- ✰ To another tube add 1 ml of test solution and 9 ml of chloroform and read (A).
- ✰ To a tube containing 1 ml of chloroform extract of sample, add 9 ml of Antimony trichloride reagent rapidly from a blow out pipette and measure the maximal blue color as indicated by the full swing of the galvanometer usually attained within 4 seconds (B).

☆ To another tube containing 1 ml of Vitamin A standard solution. Add 3 μg of Vitamin A with the help of micropipette. Followed by add 9 ml of the Antimony trichloride reagent and read the maximal color (C) at 660 nm

$$\mu g \text{ of Vitamin A}/100g = \frac{B - A \times 3 \times \text{dilution factor} \times 100}{C - A}$$

3. Estimation of Vitamin E

AOAC Official Method 992.03 Vitamin-E Activity (All rac-α-Tocopherol)

Principle

Vitamin-E activity in food samples is determined by saponification of all-rac-α-Tocopherol with organic solvent, separation from sample matrix, and quantification by liquid chromatography.

Apparatus

Partitioning Apparatus

1. Liquid chromatography (LC)-capable of pressures upto 3000 psi, with injector capable of 100 μl injetions. Operating conditions: eluent flow rate 2.0± 0.2ml/min; temperature ambient.
2. Detection-capable of measuring absorbance at 280 nm, with sensitivity 0.02 AUFS.
3. Precolumn-2mm lid × 2cm stainless steel, plackd with 40 mm pellicular reversed phase C_{18} (Alltech 28551 is suitable)
4. Column-4.6 mm id × 25 cm stainless steel, packed with 5 μm silica (Hypersil silica is suitable).
5. Shaking water bath-capable of maintaining 70±2° C, with variable speed capable of 60 oscillations/min, ca 11x14" sample area (Precision Scientific model is suitable).
6. Glassware- (i) 125ml separatory funnels (ii) 5ml volumetric flasks. (iii) 100ml low-actinic volumetric flask.

Regents

1. *Mobile phase solution*: Hexane-iso propyl alcohol (99.92+0.08, volume/volume), HPLC grade solvents. Degas 2-5 min under vacuum
2. *Wash solution*: H_2O-absolute ethanol (3+2; volume/volume
3. *Extraction solution*: Hexane-methylene chloride (3+1, vol./ vol.), HPLC grade solvents.
4. *Saponification solution*: 10.5N potassium hydroxide (KOH). Dissolve 673 KOH in 1L H_2O.
5. *Antioxidant solution*: 1 per cent pyrogallol. Dissolve 5.0g pygrogallol (1,3,5-trihydroxy benzene, 98 per cent ; Aldrich is suitable source) in 500ml absolute ethanol.
6. *Standard solutions*: (1) *Stock standard solution*: 0.5 mg/ml all- rac-α-Tocopherol acetate in hexane. Accurately weigh ca. 50mg all- rac-α-Tocopheryl acetate in hexane. (USP reference standard) into 100 ml low-actinic volumetric flask and dilute to volume with hexane (HPLC) grade. Shake well to dissolve make fresh every 3 weeks. Store al– 20°C in explosion proof freezer when not in use. (2) *Working standard solution*: 10 µg/ml all rac-α-Tocopheryl acetate. Pipet 2ml stock standard solution, (1) into 100 ml low-actinic volumetric flask. Evaporate to dryness under nitrogen. Dissolve residue in antioxidant solution (e) and dilute to volume. Prepare fresh daily
7. *Suitability test solution*: Approximately 15 µg/ml all- rac-α-Tocopherol (USP reference standard) and all- rac-α-Tocopherol acetate (USP reference standard) in hexane (HPLC grade)

Procedure

Extraction of Standard and Samples

Pipet 10.0ml working std. solution (f) (2), or sample volume containing ca 0.095 IU Vit-E activity 110 ml for ready-to-feed formulas) into 150ml centrifuge tube. Bring sample volume to 10ml with H_2O, if necessary. To standard tubes add 10ml H_2O, 20ml antioxidant solution (e) and 5ml specification solution (d). To sample tubes add 30ml antioxidant solution and 5ml saponification solution. Cap tubes and swirl briefly to mix. Place tubes in 70°C shaking H_2O bath (ca. 60 oscillations/min) for 25 min. Remove tubes and place in ice 5 min, or until contents cool to room temperature.

Quantitatively transfer contents to separate 125ml separatory funnels. Wash remaining sample or standard from tube into funnel with 30ml H_2O. Pipet 30.0 ml extraction solvent C (c) into funnel and shake ca. 2min. When layers separate, discard aqueous (lower) layer. Add some wash solution (b) to funnel and shake very gently 30s, venting frequently. Let phases separate and discard aqueous layer. Repeat wash step 3 X. Pipet 20.0ml portion from funnel into 50ml tube and evaporate to dryness under nitrogen. Transfer residues quantitatively to separate 5ml volumetric flasks and dilute to volume with mobile phase solution (a). Inject 100 µl standard or sample into LC.

System Suitability Test

Inject 100µl test solution (g) into LC. Typical peak retention times for tocopherol and tocopheryl acetate are 6.0 and 5.0 min, respt. Calculate resolution (R) factor between tocopherol and tocopheryl acetate as follows:

$$R=2\,(t_2-t_1)/(W_1+W_2)$$

where, t_1 and t_2 = retenteontune measured from injection time to elution time of peak maximum of tocopherol and tocopheryl acetate, respt., and W_1 and W_2=width of peak measured by extrapolating relatively straight sides to baseline of alcohol and acetate respt. If R factor is >1.0, proceed with sample analysis. If R factor is < 1.0; decrease amount of isopropyl alcohol added/litre (mobile phase solution) (a) by ca 0.01 per cent. Inject working standard solution (f) (2) 5x. Calculate reproducibility of replicate injection in terms of standard deviations (per USP), which should be < 2 per cent. Typical relative std. deviation values for peak height are ± 1.5 per cent.

Liquid Chromatography

Inject 100ml standard or sample into LC.

Calculations

Measure peak heights or peak areas of all- rac-α-Tocopherol in both sample and standard chromatograms. Calculate IU per reconstituted quart of Vt.-E activity (A) as follows:

$$A = H^{sam}/H^{std} \times Cstd \times 0.001\ IU/\mu g \times 946.33\ ml/quart$$

where,

Hsam = Peak height of sample.

Hstd = Peak height of standard.

Cstd. = Concentration of standard, µg/ml

Ref: J. AOAC Ent. 76, 398 (1993).

4. Estimation of Thiamine (Vitamin B_1) (Fluoremetric Method–AOAC)

Principle

This method depends upon the alkaline oxidation of Thiamine to thiochrome, which exhibits an intense blue color Fluorescence and is measured fluoremetrically.

Reagents

1. *Acid KCl solution*: Dissolve 125 g KCl in H_2O to it. Add 4.3 ml concentrated HCl and dilute to 500 ml with H_2O.
2. *15 per cent Sodium hydroxide solution*: Dissolve 37.5 g of NaOH in water to make 250 ml.
3. *1 per cent Potassium ferricyanide solution*: Dissolve one gram $K_3Fe(CN)_6$ in H_2O to make 100 ml. It should be prepared on the day it is used.
4. *Oxidizing reagent*: Mix 4 ml of 1 per cent $K_3Fe\ (CN)_6$ with the 15 per cent NaOH solution to make 100 ml. Use within 4 hours.
5. *Iso-butyl alcohol*: Redistill in glass apparatus.
6. *Quinine Sulphate stock solution*: It is to govern the reproducibility of fluoremeter. Dissolve 10g quinine sulfate in 0.1 H_2SO_4 to make 1 litre. Store in amber colored bottle.
7. *Quinine sulfate standard solution*: Dilute 1 ml of quinine sulfate stock with 39 vol. of 0.1 H_2SO_4. This solution fluoresces to the same degree as does iso-butanol extraction of thiochrome obtained from 1µg Thiamine HCl. Store in amber colored bottle.
8. *Thiamine HCl standard solution*:
   (*i*) *Stock solution*: Accurately weigh 50 mg of (well dried over P_2O_4, in a dessicator) Thiamine hydrochloride.

Dissolve in 20 per cent ethanol, adjust pH 3.5–4.3 with HCl and dilute to 500 ml with the acidified alcohol (100 µg/ml).

(*ii*) *Intermediate solution*: (10 µg/ml) Dilute 10 ml solution to 100 ml with acidified alcohol. Store at 10°C in a colored bottle.

(*iii*) *Working solution*: (1 µg/ml)–To 10 ml intermediate add 50 ml 0.1 N HCl and digest in autoclave for 30 min. under 15 lb pressure. Then dilute to 100 ml with 0.1 N HCl to 100 ml.

9. *2 N Sodium acetate solution*: Dissolve 68 g sodium acetate trihydrate in H_2O to make final volume 250 ml.
10. *Bromo cresol green pH indicator*: Dissolve 0.1g indicator by triturating in agate mortar with 2.8 ml 0.05 N NaOH and dilute to 200 ml with H_2O. Transition range 4.0 (green)–5.3 (blue).
11. *Bromo-phenol blue indicator*: Dissolve 0.1 g indicator by triturating in agate mortar with 3 ml 0.05 N NaOH and dilute to 250 ml with H_2O Transition range 3.0 (yellow)–4.6 (blue).
12. *Enzyme solution*: Prepare 10 per cent Ag solution of enzyme preparation potent in distatic and phospholytic activity (Mylaso, clarase).
13. *Base-exchange silicate*: Weigh 100g of Base Exchange silicate in beaker, add enough hot 3 per cent acetic acid to cover and boil for 10-1.5 minute, stirring continuously. Let settle and decant supernatant. Repeat washing 3 times. Then wash three times with hot KCl. Finally wash with H_2O till Cl free. Dry at 100°C and store in a well-closed container.
14. *Chromatographic tubes*: Use chromatographic tubes (275 mm length, 60 ml reservoir) with reservoir at top, absorption tube in the middle and capillary tube at the bottom. Over upper end of capillary, place fun glass wool. To the middle tube add water suspension of base-exchange silicate. Wash all silicate from wall of reservoir. A layer of liquid should be present above silicate.

Procedure

Extraction (For Sample Hydrolysis)

- ☆ Weigh 2.5g of dried and well-ground sample and add 25 ml 0.1N HCl. Shake well.
- ☆ Wash sample from the sides of flask.
- ☆ Autoclave for 30 min. under 15 lb pressure.
- ☆ Cool and filter. Dilute to 100 ml with 0.1 N HCl.

Enzymatic Hydrolysis (For Sample Containing Thiamine Pyrophosphate)

- ☆ Weigh 2.5 g of dried and well-ground sample (10g for semi-dry sample)
- ☆ Add 50 ml 0.1 N HCl. Autoclave, adjust pH to 4.0-4.5 with sodium acetate solution.
- ☆ Add 30.5 ml enzyme solution, mix and incubate for 3h at 45-50°C.
- ☆ Cool, adjust pH 3.5 dilute to 100 ml with H_2O.
- ☆ Filter through ash free filter paper.

Purification

- ☆ Pass aliquot of filtered solution through prepared chromatographic column
- ☆ Wash column with 5 ml portion of hot water. Elute thiamine by passing five 4 ml portions of boiling acid KCl through column.
- ☆ In the meantime, no air should pass through the column.
- ☆ Collect elute in 25 ml volumetric flask and dilute to volume with acid KCl solution.

Oxidaiton of Thiamine to Thiochrome

- ☆ To 4 tubes (capacity 40 ml) add 1.5 g NaCl or KCl and 5 ml working standard solution and mix.
- ☆ To two tubes add 3 ml oxidizing reagent rapidly and shake vigorously. To the remaining two tubes add 3 ml of 15 per cent NaOH solution similarly (Blank).

- ☆ To two tubes add 5 ml sample aliquot and 1.5 g NaCl or KCl.
- ☆ To remaining two tubes, add 3 ml oxidizing reagent and to other two tubes add 3 ml 15 per cent NaOH (Blank).
- ☆ To all tubes add 13 ml Iso-butanol and shake over whirl-mixer for two min.
- ☆ Pipette 19 ml of clear supernatant and measure fluorescence of each tube with input T range 365 nm and output T range 435 nm.
- ☆ Use standard Quinine Sulphate to govern the reproducibility of fluoremeter.

Calculations

W = Net weight (g) of sample

V = Volume of aliquot made

S = Reading of standard

d = Reading of standard blank

I = Reading of sample

b = Reading of sample blank

1 µg/ml = Concentration of Standard Thiamine working Solution

$$\text{Thiamine mg/100g} = \frac{(I - b) \times 1 \times V \times 100}{(S - d) \times 1000 \times W}$$

5. Estimation of Riboflavin (Fluoremetric Method)

Principle

Riboflavin is extracted with hot dilute acid. After adjusting pH to 4.5, the extract is filtered and decolorized with potassium permanganate and hydrogen peroxide. Then the fluorescence of the solution is measured before and after reduction with dithionite solution. The difference in fluorescence being proportional to riboflavin contents.

Reagents

1. *0.1 N HCl*: Dilute 4.5 ml of concentration HCl to 500 ml with H_2O.

2. *4 per cent $KMnO_4$*: Dissolve 4g of $KMnO_4$ in water to make 100 ml.
3. *3 per cent H_2O_2*: Dilute 30 per cent H_2O_2 to 100 ml with water.
4. *0.02 N Acetic acid*: Dilute 1.1 ml glacial acetic acid to 1 litre with water.
5. *Riboflavin standard*: (100µg/ml)
   (*i*) *Stock Solution*. Dissolve 50 mg of Riboflavin previously dried and stored in dark in desiccator. Dissolve in sufficient quantity of 0.002 N acetic acid on steam bath with constant stirring. Cool and add 0.02 N acetic acid to make volume 500 ml.
   (*ii*) *Intermediate solution*: (10 µg/ml). Dilute 10 ml of stock solution to 100 ml with 0.02 N CH_2 COOH. Store under dark at 10°C.
   (*iii*) *Working solution*: (1 µg/ml) Dilute 10 ml intermediate solution to 1 litre with water avoiding exposure to strong light.
   (*iv*) *Sodium hydrosulfite Na_2SO_6*: $2H_2O$.

Procedure

(a) Extraction

- ✰ Weigh 2.5g sample (estimated to contain-not less than 10 µg of riboflavin) into a 100 ml conical flask and add 25 ml 0.1 N HCl. Cover the flask with inverted beaker.
- ✰ Autoclave the sample at 15 lb for 30 min.
- ✰ Cool the sample, wash into a 100 ml beaker.
- ✰ Set pH of the sample solution to 6.8 with 40 per cent NaOH
- ✰ Now add immediately HCl drop wise, whilst stirring to bring pH 4.5.
- ✰ Transfer the extract to a 100 ml volumetric flask and dilute to 100 ml with water.
- ✰ Filter the extract with a filter paper.

(b) Decolorizing the Sample

- ✰ Pipette 10 ml aliquots of the sample solution into four 25 ml test tubes.

- ☆ Add 2 ml standard riboflavin working solution (0.1 µg/ml) to two tubes.
- ☆ Add 2 ml H_2O to the other two tubes.
- ☆ Add the glacial acetic acid to each tube and mix.
- ☆ Add 0.5 ml of 4 per cent $KMnO_4$ and mix.
- ☆ After 2 min add 0.5 ml 3 per cent H_2O_2.

(c) Measurement of Fluorescence

- ☆ Adjust the fluoremeter to give a deflection of 0 against 0.1 N H_2SO_4 and 100 against the sample positive standard solution using wavelength 525 nm.
  - § Primary filter No. 621
  - § Secondary filter No. 4015
- ☆ Measure the fluorescence of each tube in turn.
- ☆ Then add 20 mg of Sodium dithionite to each tube in turn and mix. Read the blank value within 10 sec.

Calculations

W = Net weight of sample
100 ml = Volume of extract
10 ml = Aliquot taken for decolorizing
T = Reading of sample positive standard
S = Reading of sample
S_b = Reading of sample blank
Tb = Reading of sample positive standard blank

Hence sample positive standard florescence = T – Tb

Sample florescence (0.2 µg) = (T–Tb)-(S-S_b).

Riboflavin content (µg/100 g) of sample =

$$\frac{S - Sb}{(T - Tb) - (S - Sb)} \times \frac{0.2}{W}$$

6. Estimation of Vitamin C (Ascorbic Acid)

Method–I: (2, 6 Dichloroindophenol Method, AOAC)

Application

The method is applicable to foodstuffs that give colorless or faintly colored extracts.

Principle

Ascorbic acid reduces oxidation-reduction dye, 2, 6 dichloroindophenol to colorless solution. At end point, excess unreduced dye is rose to pink in acid solution. Vitamin is extracted and titration performed in the presence of HPO_3-CH_3COOH solution to maintain proper acidity for reaction and to avoid autoxidation of ascorbic acid till high pH.

Reagents

1. *Metaphosphoric acid-acetic acid solution*: Dissolve 15 g HPO_3 pellet in 40 ml acetic acid and 200 ml H_2O, dilute to 500 ml. Filter rapidly into a colored bottle.
2. *Ascorbic acid standard solution*: 1 mg/ml. Weigh 50 mg Ascorbic acid standard that has been stored in dessicator away from sunlight. Transfer to 50 ml volumetric flask and dilute with HPO_3–CH_3COOH solution.
3. *Indophenol standard solution*: Dissolve 50 mg 2,6 dichloroindophenol that has been stored in dessicator away from sunlight over soda lime, in 50 ml H_2O and add 42 mg $NaHCO_3$. Shake vigorously and dilute to 200 ml with H_2O, Filter and store in an amber glass bottle.

Procedure

Extraction

(a) For dry Materials

- ☆ Weigh 5g sample and add 50 ml HPO_3–CH_3COOH solution.
- ☆ Grind the sample gently in a mortar. Adjust pH to 1.2.
- ☆ Triturate the sample until it is in suspension.
- ☆ Filter rapidly through Whatman No. 1 and dilute to 100 ml with metaphosphoric acid reagent.

(b) For Fruit and Vegetable Juices

- ☆ Extract juice of fruits by pressing or squeezing well-pulped fruit and filter rapidly.
- ☆ Add extracted juice to equal volume of HPO_3–CH_3COOH and designate total volume V.

Determination

- ☆ Transfer three 2.0 ml aliquots ascorbic acid standard to each of 50 ml conical flasks containing 5 ml HPO_3–CH_3COOH solution.
- ☆ Titrate rapidly with indophenol solution until light but distinct rose pink color persists for more than 5 sec. (Generally 15 ml dye solution is required for it).
- ☆ Similarly take 7 ml of HPO_3–CH_3COOH solution in a conical flask. Add H_2O equal to volume of indophenol used against standard and titrate for blank. Take three readings.
- ☆ Take 2 ml sample aliquot in a conical flask containing 5 ml HPO_3–CH_3COOH solution. Titrate against the dye solution. Take three readings similarly.

Calculations

W = Net weight of sample

V = Volume of aliquot made

X = Volume of dye solution used against standard

Y = Volume of dye solution used against sample aliquot

B = Volume of dye solution used against blank

Ascorbic acid (mg/100 g) =

$$\frac{Y-B}{X-B} \times \frac{1 \times V}{X-B} \times 100$$

Method–II: Estimation of Ascorbic Acid in a Given Food Sample

In this method ascorbic acid is just oxidized to dehydroascorbic acid. It reacts with 2,4 dinitro phenylhydrazine, which is dissolved in the sulfuric acid and absorption manimum is read at 500-520 nm.

Reagents and Materials

1. *Metaphosphoric acetic acid solution*: Dissolve 50 g of metaphosphoric acid, add to it 100 ml of acetic acid make up to 1 litre.
2. *2.4 dinitro phenyl hydrazine*: Dissolve 2 g of crystalline compound in 100 ml of 9N H_2SO_4. Filter and store in brown bottle in refrigerator. Refilter a portion of it before use.
3. *Thiourea solution*: Dissolve 10 g thiourea in 100 ml of 50 per cent ethanol and refrigerated.
4. *Cupric Sulfate solution*: Dissolve 1.5g of cupric sulfate in water and dilute it to 100 ml.
5. *Combined color reagent*: Prepare fresh on the day of use by mixing 15 ml of 2, 4 DNPH. 20.1 ml of cupric sulfate solution and 30.1 ml of thiourea.
6. *85 per cent H_2SO_4*: To 30 ml of distilled water add 170 ml of conc. H_2SO_4 mix cool stored in glass stoppered bottle.
7. *Standard solution*: Prepare fresh stock of Ascorbic Acid by dissolve 100 mg. in 100 ml of metaphosphoric acid. Dilute 5 ml of stock to 100 ml. 1 ml of diluted solution is equal to 50µg of Ascorbic Acid.

Procedure

- ✰ Dissolve weighed amount of sample in 100 ml of metaphosphoric acid and filter.
- ✰ Pipette 1 ml of extract into small test tubes run the experiment in duplicate.
- ✰ To each test tube add 1 ml of metaphosphoric acid except blank, in which 2 ml metaphosphoric acid is taken
- ✰ To each test tube, take 0.4 ml of freshly prepared color reagent and mix.
- ✰ Cool in the ice bath for 5 min.
- ✰ Add slowly 2 ml of ice cold 85 per cent H_2SO_4 and mix.
- ✰ Let the tubes stand at room temperature for 30 min then mix
- ✰ Transfer the sample to cuvettes with not exceeding 3 ml, measure the absorbance against blank at 500 nm.

☆ Stand the solution in the concentration of 10, 20, 30, 40 and 50 μg which are treated in the same way, standard curve is drawn and ascorbic acid content is estimated.

3. Enzymes

Separation of Alkaline Phosphatase from goat intestine by ion exchange chromatography

Principle

The principal feature underlying in ion exchange chromatography is the attraction between oppositely charged particles. Many biological materials, for example amino acids and proteins have ionisable groups and the fact that they may carry a net positive or negative charge can be utilized in separating mixtures of such compounds. The net charge exhibited by such compounds is dependent on their PKa value and on the pH of the solution in accordance with the Hendersen-Hasselbach equation.

Ion-exchange separations are mainly carried out in columns packed with an anion exchanger or a cation exchanger. Cation exchanger possess negatively charged groups and these will attract positively charged molecules, whilst an anion exchangers have positively charged groups which will attract negatively charged molecules.

Diethyl Amino ethyl-Agarose (DEAE-Agarose) used in this experiment contain positively charged groups in the resin matrix, so is an anion exchanger.

The molecules of alkaline Phosphatase are glycoproteins containing various amounts of sialic acid. These forms therefore carry different negative charges and this property is used to separate them on an ion exchange column.

Reagents and Materials

1. Goat intestine
2. Ice
3. Normal saline
4. 10 per cent homogenate in 25 mM sucrose
5. 40-60 per cent Ammonium Sulphate

6. 0.1M Glycine NaOH buffer pH 9.0
7. 60 mM PNPP
8. 60mM $ZnCl_2$
9. 0.5M phosphate pH 7.0
10. DEAE Sepharose column

Procedure

Preparation of Sample Homogenate

- Goat intestine pieces kept in ice, were cut into small pieces in a beaker
- Normal saline solution containing 10 per cent homogenate in 25 mM sucrose solution was added to the intestinal pieces
- 5g goat intestine cut into pieces with the precaution to keep the pieces on ice.
- Grind in mortar and pestle
- Prepare 10 per cent homogenate in 25 mM sucrose solution; the final volume made is 50 ml.
- Centrifuged at 1000 rpm for 20 min, Nuclei and unbroken cells will settle down and got cell homogenate
- Assay the enzyme in supernatant *i.e.* cell homogenate
- Take cell homogenate 0.1-0.2 ml
- Prepare 0.1 M Glycine-NaOH buffer pH 9.0.
- 1.8 to 1.9 ml buffer added with 0.1 to 0.2 ml of cell homogenate
- Took 60 mM PNPP 0.2 ml and 60mM $ZnCl_2$ 0.2 ml are added to the above cell homogenate
- Incubated the whole thing at 37°C for 30 min
- Added 5ml of 0.1M NaOH and OD read at 405 nm

Salting out or Ammonium Sulphate Precipitate

1st Step

- Took cell fractions, added 0-40 per cent of $(NH_4)_2SO_4$ salt, kept at 4°C for 3-4h, centrifuged at 5000rpm for 2 min.

- ☆ Supernatant precipitated with 40 to 60 per cent $(NH_4)_2SO_4$ solution and kept for 2h in a refrigerator
- ☆ Precipitate of first step dissolved in buffer 0.5 M phosphate pH 7.0 and dialyzed overnight to get rid of Ammonium Sulphate

2nd Step

- ☆ Centrifuged- precipitate and supernatant for 2nd step. Calculated the units could be lesser than the total units
- ☆ Took supernatant and proceeded with 60 to 90 per cent Ammonium salt

Chromatography Column and Elution

- ☆ Prepared DEAE-sepharose column and equilibrated with phosphate buffer
- ☆ Sample was applied on top of the column and eluted
- ☆ Collected the unbound fraction and read absorbance at 280nm
- ☆ One peak is obtained.

Chapter 5
Food Additives and Preservatives

1. Estimation of Nitrite in Meat and Meat Products

Method–I

Reagents and Materials

1. Hot alkaline water
2. Sulfanilic acid solution
3. α-napthylamine
4. Glacial acetic acid
5. Standard sodium nitrite solution
6. Saturated mercuric chloride solution
7. 500 ml Erlenmeyer flasks
8. Spectrophotometer
9. 50 ml volumetric flasks
10. Pipette
11. Oven

Procedure

Preparation of Sample

- ☆ Take 20 g of meat sample and mix thoroughly using a small amount of hot alkaline water.

- Transfer the contents to a 500 ml Erlenmeyer flask
- Wash the container and cover with several portions of the hot water, adding all washing to the flask.
- Add enough hot alkaline water to bring the volume to approximately 300 ml.
- Place the flask with a stopper in an oven at 80°C for 2 h, shaking occasionally.
- Add 5 ml of a saturated mercuric chloride solution mix and cool at room temperature.
- Transfer the entire contents to a 500 ml volumetric flask and bring to volume with water and shake well.
- Filter a portion through No. 42 Whatman filter paper into a test-tube.

Determination of Nitrite

- Pipette 2 ml aliquot of the filtered solution into a 50 ml volumetric flask.
- Add 1 ml of sulfanilic acid and 1 ml of μ–napthylamine. Fill the flask to the 50 ml volume mark with water.
- Mix well and let it stand for 1 h to develop the red color.
- Prepare a blank containing the reagents only and water.
- Transfer a portion of solution to photometer cell and determine the absorbance at 520 nm, setting instrument to zero absorbance with the blank.

Calculations

$$\text{Sodium nitrite (ppm)} = \frac{C \times 200 \times V}{W \times V_1}$$

where,

W = Weight of sample taken (g)

V = Volume made (ml)

V_1 = Volume of aliquot taken for estimation (ml)

C = Concentration of sodium nitrite from standard curve (ppm).

Observations

Sample No.	Weight of Sample W	Volume Made V	Aliquot Taken V_1	Conc. from Standard Curve C	Sodium Nitrite (ppm)
1.					
2.					
3.					

Method–II: Estimation of Nitrites (AOAC, 1975)

Reagents and Materials

1. *Modified Griess reagent*: Dissolve 0.5 g sulphanilic acid in 150 ml 15 per cent (V/V) acetic acid. Boil 0.1 per cent L-naphthylanine or 0.125 g of the hydrochloride in 20 ml water until dissolved and pour while hot into 150 ml acetic acid. Mix the two solutions, filter if necessary and store in brown glass bottle.
2. *Nitrite standard solution*: 0.1 and per ml dissolve 1.1 g $AgNO_3$ in nitrite free water, precipitate Ag with NaCl (about 1 g) and shake until AgCl floccolates. Dilute to 1 litre, mix and settle. Dilute 100 ml to 1 litre and 10 ml to 1 litre using nitrite free water in each case.
3. Volumetric Flask
4. Spectrophotometer.

Procedure

- ☆ Weigh 5.0 g finely comminuted and thoroughly mixed sample into 50 ml beaker.
- ☆ Add about 40 ml nitrite from water heated to 80°C. Mix thoroughly with glass rod, taking care to break up all lumps and transfer to 500 ml volumetric flask.
- ☆ Thoroughly wash beaker and rod with successive portion of the hot water adding all washings to flask.
- ☆ Add enough, hot water to bring volume to about 300 ml.
- ☆ Transfer flask to steam bath and let it stand for 2 hours shaking occasionally.

- ☆ Add 5.0 ml saturated $HgCl_2$ solution and mix.
- ☆ Cool to room temperature, dilute to volume with nitrite free water and mix again.
- ☆ Filter, dilute suitable aliquot to volume in 50 ml volumetric flask, add 2 ml reagent, mix and let color develop (1 hour).
- ☆ Transfer suitable portion of solution to photometer cell and determine absorbance at 520 nm against blank of 50 ml water 2 ml reagent.
- ☆ Determine nitrite present by comparison with standard curve prepared as follow; dilute suitable volume standard nitrite solution to volume in 50 ml volumetric flask, add 2 ml reagent and proceed as above. Standard curve is straight line to 5 µg N in final solution.

2. Estimation of Nitrate in Meat and Meat Product

Determination of Nitrates (AOAC, 1975)

Reagents and Materials

1. m-Xylenol
2. 2, 4-Dimethylphenol
3. *Silver-ammonium hydroxide solution*: Dissolve 5 g nitrate free Ag_2SO_4 in 60 ml NH_4OH. Heat to boiling point, concentrated to about 30 ml, cool, and dilute to 100 ml with water.
4. *Bromocresol green indicator*: Dissolve 0.1 g bromocresol green in 1.5 ml N/10 NaOH and dilute to 100 ml with water.
5. *Nitrate standard solution*: Dissolve 0.1805 g recrystalised KNO_3 in water and dilute to 1 litre or dilute 17.85 ml 0.1 N HNO_3 to 1 litre, 10 ml contains 0.25 mg nitrate.

Procedure

- ☆ Mix 5 to 10 g finely comminuted and thoroughly mixed sample with 80 ml warm water.
- ☆ Break up all lumps and heat on steam bath for 1 hour, stirring occasionally.

- ☆ Transfer to 100 ml volumetric flask, cool, dilute to volume and mix.
- ☆ Filter or let it settle, and pipette 40 ml filtrate or supernatant, in 250 ml volumetric flask (no correction for volume occupied by meat is necessary).
- ☆ Add 3 drops bromocresol green indicator. Add H_2SO_4 (1:10) drop wise until color changes to yellow.
- ☆ Oxidise nitrites to nitrate by adding 0.2 N $KMnO_4$ solution drop wise with shaking until faint pink color remains about 1 minute. Add 1 ml H_2SO_4 (1:10) and 1 ml phosphotungstic acid solution (20 g/100 ml).
- ☆ Dilute to volume to 250 ml, mix and filter.
- ☆ Into 500 ml flask measure aliquot (?20 ml) containing 0.025 to 0.25 mg nitrate N.
- ☆ Add enough $AgNH_4OH$ solution to precipitate chlorides and most of excess phosphotungstic acid (slight excess of Ag reagent is not harmful, 1 or 2 ml is usually enough).
- ☆ Without decanting or filtering add volume H_2SO_4 (3+1) about 3 times volume liquid in flask.
- ☆ Put stopper on the flask, mix, cool to about 35°C and add 0.05 ml (1-2 drops) of the m-xylenol, shake and held for 30 minutes at 30-40°C (Yellow to brownish yellow color, indicative of nitrates, appears. Bright red precipitate, due to incomplete phosphotungstic acid may also appear. Slight excess of phosphotungstic acid possess no interference but large excess may do so).
- ☆ After nitration is complete, add 150 ml water, taking care to wash off stopper, and distill 40-50 ml into receiver containing 5 ml NaOH (10 g/lit).
- ☆ Quickly remove any nitroxylenol solidifying in condensor by stopping water flow and letting condensor becomes warm.
- ☆ Transfer distillate to 100 ml volumetric flask.
- ☆ Dilute to volume with water, and determine nitrate N by comparing reading of color of suitable aliquot with standard curve prepared at about 450 nm.

- Prepare color standard from 10 ml nitrate standard solution, using 0.05 ml m-Xylenol and 30 ml H_2SO_4 (3+1) and dilute distillate to 500 ml.

Determination of Nitrate

- Pipette 20 ml aliquot into reduction column that has just been washed with 10 ml of double distilled water. Column should not be allowed to go dry.
- For high protein meat products, make a 10 times dilution and place 20 ml of the diluted solution on the column.
- Wash column with 15 ml NaCl solution followed by 10 ml double distilled water and collect total eluant in 100 ml volumetric flask. Rinse column with 10 ml double distilled water between samples.
- Add 5 ml sulfanilamide to flask, mix well, let stand for 3 min. Add 2 ml coupling reagent, mix and dilute to 100 ml, with double distilled water. Allow color to develop at least 20 min.
- Read absorbance at 540 nm against a reagent blank that has been carried through the entire procedure and color developed with sulfanilamide and coupling reagent Developed color is stable at least for 2h
- Preparation of standard curve: Place 0, 1, 2, 3, 4 and 5 ml of nitrate standard into separate 100 ml beakers. Add 5 ml of buffer solution to each standard and dilute to about 20 ml. Run standards through column and wash as with samples.
- Collect eluent in 100 ml volumetric flask and develop color as with samples
- Read absorbance at 540 nm.

Precaution

- For every new column or color development solution, a new standard curve is run.

Observations

Sample No.	*Weight of Sample*	*Volume of Extract*	*Aliquot Taken*	*Conc. of Nitrate from Standard Curve*	*Burrette Concen-tration*	*Sodium Nitrate*
	W	*V*	*V1*	*C*	*A*	*(ppm)*
1.						
2.						
3.						

Calculations

$$\text{Sodium Nitrate (ppm)} = \left(C \times \frac{V}{V_1} \times \frac{100}{W} \times 11.08 \right) - (A \times 1.348)$$

where,

W = Weight of sample taken

V = Volume of extract made

V_1 = Aliquot taken for estimation.

C = Concentration of nitrate from standard curve.

A = Nitrite concentration (ppm) in sample.

3. Estimation of Phosphate Content

Method–I

Reagents and Materials

1. Diluted nitric acid
2. Quimociac reagent
3. Gooch crucible
4. Glass fibre
5. Filter paper
6. Steam bath
7. Oven
8. Dessicator

Procedure

- ☆ Weigh 2.5 g of sample in an ashing dish.
- ☆ Dry for 30 minutes at 125°C in a forced draft oven.
- ☆ Ash at 55°C
- ☆ Add 25 ml diluted nitric acid.
- ☆ Heat on steam bath for 30 min
- ☆ Filter into a 400 ml beaker
- ☆ Wash dish and paper with dry weight and make the volume 100 ml.
- ☆ Run a reagent blank in parallel, using 25 ml diluted nitric acid and 75 ml dry weight.
- ☆ Add 50 ml quimociac reagent to test beaker, cover with a watch glass, and boil for 1 min (do not use an open flame).
- ☆ Cool to room temperature while swirling carefully.
- ☆ Transfer the precipitate to the prepared crucible and wash 5 times with 25 ml portions of distilled water.
- ☆ Allow each portion to drain thoroughly before adding next portion.
- ☆ Dry the crucible and contents at 250°C for 30 min.
- ☆ Cool in a dessicator and weigh.

Observations

Sample Number	*Weight of*			*Phosphorus (Per cent)*
	Sample W	*Blank B*	*Precipitate A*	
1				
2				
3				

Calculations

$$\text{Phosphorous (per cent)} = \frac{[(A - B) \times 0.014 \times 1000)]}{W - (0.0106 \times \text{Per cent meat protein})}$$

where,

A = Weight of precipitate (g)

B = Weight of blank (g)

W = Sample weight (g)

0.014 = Gravimetric factor derived from atomic weight of phosphorus = 30.97 divided by the molecular weight of Quinolimium phosphomolybdate (QPM) = 2212.71

0.0106= Factor to correct for the-phosphorus content of meat protein.

Method–II: Estimation of residual polyphosphate

Total residual phosphorus was estimated in cooked meat samples according to the method of Fiske and Subbarow (1925)

Reagents and Materials

1. Concentrated Hydrochloride
2. Amino Naphthol Sulphonic Acid (ANSA)
3. Potassium dihydrogen phosphate
4. Burner
5. Volumetric flask
6. Spectrophotometer

Procedure

- ✰ Fifty mg of dried cooked meat was digested in corning wide mouth test tube on a burner with 5 ml nitric acid: Perchloric acid (10:1) till all acid got evaporated;
- ✰ Added 1ml of cone. HCl and again evaporated, cooled down.
- ✰ Added 1 per cent HCl to make volume of 25 ml in the volumetric flask.
- ✰ To 2.0 ml of this digested sample, added 2.2 ml distilled water (making the volume to 4.2 ml), 0.2ml of 10N Sulphuric acid, 0.4 ml of 2.5 per cent aqueous solution of Ammonium molybdate and 0.2 ml of Amino naphthol sulphonic acid*, color was allowed to develop for 10 min. and transmittancy was read at 660 nm in Spectrophotometer.

☆ Standard curve was prepared using a series of standard phosphorus solutions. For this dissolved 1.099 g Potassium dihydrogen phosphate (KH_2PO_4) in 250ml distilled water, which gave standard phosphorus solution containing 1mg phosphorus per ml of solution. One ml of this solution was further diluted to make 100 ml in a volumetric flask with distilled water to give 10µg phosphorus per ml of solution. Standard curve was plotted for 1to10µg phosphorus on the basis of optical density (OD). The color was developed and measured in the same way as described previously. A blank without any phosphorus was also prepared.

The amount of phosphorus in unknown samples was calculated on the basis of standard curve by following formula:

$$\text{Conc. of Unkown sample} = \frac{\text{Conc. of Standard} \times \text{O.D. of unknown sample}}{\text{O.D. of known sample}}$$

☆ In case of treated samples residual phosphorus was calculated by substracting total phosphorus of control sample from total FPP

Phosphorus of Treated Samples

Conversion of phosphorus to polyphosphate was done on the basis of mol wt. of polyphosphates.

Percent polyphosphate recovered was calculated by following formula:

$$\text{\% Polyphosphate recovered} = \frac{\text{mg Polyphosphate recovered}}{\text{mg Polyphosphate added}} \times 100$$

Note

For preparation of this solution 0.25g of Aminonaphthol sulphonic acid (ANSA) powder was mixed with 97ml of 15 per cent Sodium sulphite solution. After this the remaining ANS was dissolved by adding 20 per cent Sodium sulphite solution (2.5ml). The solution was then filtered through Whatman filter paper No. 1.

The solution was kept in dark bottles in refrigerator. It was never used after a week's storage.

4. Estimation of Soybean Flour and Soybean Protein Concentrate

Reagents and Materials

1. Ethanol 95 per cent
2. Alcohol potassium hydroxide solution (8 per cent)
3. Diluted hydrochloric acid (1-3)
4. Centrifuge
5. Centrifuge Goetz tubes of 100 ml

Procedure

- ☆ Weigh 10 g of sample in a 100 ml Goetz tube.
- ☆ Add 50 ml of 8 per cent alcoholic potassium hydroxide solution.
- ☆ Digest in a steam bath for 30 min.
- ☆ Stir it.
- ☆ Shake well and centrifuge for 4 min
- ☆ Decant and discard the supernatant solution.
- ☆ Wash residue with 25 ml of 95 per cent ethanol
- ☆ Stir the sediment thoroughly.
- ☆ Centrifuge and decant.
- ☆ Discard the alcoholic solution.
- ☆ Add 50 ml of dilute hydrochloric acid.
- ☆ Mix thoroughly, stopper, and shake for 1 min
- ☆ Centrifuge at 2000 rpm for 4 min (Retain residue for cereal determination)
- ☆ If supernatant is not clear, fitter it through a double thickness No.541 Whatman filter paper.
- ☆ Transfer 25 ml of clear supernatant to a second Goetz tube containing 75 ml of 95 per cent ethanol.
- ☆ Shake well and allow it to stand for 1 h.
- ☆ Accelerate from 0 to 1500 rpm over 1 min

- ☆ Centrifuge at 1500 rpm for exactly 2 min
- ☆ Read volume of sediment in the tube.

Observations

Sample Number	*Volume of Sediment*	*Soybean Flour (Per cent)*	*Soybean Concentrated (Per cent)*
1.			
2.			
3.			

Calculations

Soybean flour (per cent) = Volume of sediment × 6

Soy protein concentrate (per cent) = Volume of sediment × 2.5

5. Determination of Starch in Meat and Meat Products

Reagents and Materials

1. Carrez solution No.1
2. Zinc acetate solution
3. Carrez solution No.2
4. Potassium ferrocyanide solution
5. Acidified calcium chloride solution
6. Autoclave
7. Volumetric flasks (100 ml)
8. Shaker
9. Polarimeter

Procedure

- ☆ Mix about 5 g of sample with 10 ml of water in a 400 ml beaker and add 50 ml of acidified calcium chloride solution.
- ☆ Heat the mixture in an autoclave for 10 min at a pressure of 15 lb/in^2. Cool the mixture by immersion in cold water.
- ☆ Transfer it to a 100 ml volumetric flask by washing it with acidified calcium chloride solution until the volume is approximately 70 ml

- Add 2 ml of carrez solution No.1 shake the mixture well. Add 2.0 ml carrez solution No.2 Shake the mixture again. Dilute it to 100 ml at 20°C with acidified calcium solution.
- Filter the dispersion on a Whatman No.41 filter paper until the filtrate is perfectly clear.
- Discard the first 15-20 ml of filtrate and obtain the polarimeter reading at 20°C in a 20 cm tube on the subsequent running.

Observations

Sample Number	*Weight of Sample Taken*	*Starch (per cent)*
1.		
2.		
3.		

Calculations

If P is the reading at 20°C in a 20 cm tube and $[\alpha]^{20}$ is 200, then calculate the amount of starch in the sample from the following expression

$$\text{Starch (per cent)} = \frac{P \times 10^4}{400 \times \text{weight of sample taken}}$$

Precaution

- The type of cereal present affects the allowance made for the nitrogen due to the filter in the calculation of the meat content.

6. Estimation of Sulphur Dioxide in Sausages

The sulphur dioxide can be titrated directly by applying double titration method of Potter. The total reducing substances are measured by iodimetric titration of an aliquot of the filtrate obtained after macerating the sample with water, digesting with alkali and acidifying. To another aliquot hydrogen peroxide is added prior to the titration in order to oxidize sulphite to sulphate. The difference between the titrations is equivalent to the sulphur dioxide present

Reagents and Materials

1. 5N NaOH
2. 5N HCl
3. 1 per cent Starch solution
4. 0.5N Iodine
5. 3 per cent H_2O_2
6. Mechanical macerator
7. Fitter funnel
8. Magnetic stirrer
9. Titration unit.

Procedure

- ☆ Macerate 50 g of sausage sample for 1.5 min. with 250 ml of water in mechanical macerator and filter the mixture as rapidly as possible through a large fluted filter paper.
- ☆ Pipette 75 ml of filtrate into each of two flasks A and B followed by 75 ml of water and 5 ml of 5N sodium hydroxide.
- ☆ Stir each gently with a magnetic stirrer, avoiding beating air into solution and allow the solutions to stand for 20 minutes.

Titration of A: Add 7 ml of 5N hydrochloric acid and 10 ml of 1 per cent starch solution and titrate immediately with 0.05N iodine to a definite blue end point (A ml).

Titration of B: Add 7 ml of 5N hydrochloric acid, 2 ml of 3 per cent hydrogen peroxide and 10 ml of starch solution. While stirring the mixture gently, titrate it immediately with 0.05N iodine to a definite blue end point (B ml).

Precautions

- ☆ Maceration time is fairly critical and is likely to vary according to the particular blender used.
- ☆ Filtration time should be kept as short as possible.
- ☆ Method appears to be inapplicable to badly deteriorated samples.

Observations

Sample Number	*Volume of 0.05 Iodine Used*		*Sulphur Dioxide (ppm)*
	for A	*for B*	
1.			
2.			
3.			

Calculations

Sulphur dioxide (ppm) = 107 (A-B)

7. Estimation of Added Cereals in Meat and Meat Products (Barker and Summerson, 1941)

Reagents and Materials

1. 95 per cent Ethanol
2. 8 per cent Alcoholic caustic potash solution
3. Diluted hydrochloric acid (1:1)
4. Centrifuge
5. Centrifuge tube 100-ml
6. Gooch crucible.

Procedure

☆ Weigh 10 g of sample.

☆ Extract with two successive 50 ml portions of warm distilled water.

☆ Shake, centrifuge, decant and discard the supernatant liquid after each extraction.

☆ Add 50 ml of 8 per cent potassium hydroxide solution and digest in a steam bath for 20 min with occasional stirring

☆ Dilute to 100 ml with 95 per cent ethanol.

☆ Centrifuge the 100 ml suspension for 5 min.

☆ Decant and discard the supernatant liquid.

☆ Wash the residue 25 ml of 95 per cent ethanol by stirring the sediment thoroughly.

- ☆ Centrifuge, decant, and discard the supernatant liquid.
- ☆ Add 50 ml of 1:1 hydrochloric acid.
- ☆ Mix thoroughly, stopper and shake for 1 min.
- ☆ Centrifuge at 2000 rpm for 4 min.
- ☆ If supernatant liquid is not clear, filter it through double thickness No. 41 Whatman filter paper.
- ☆ Transfer 25 ml of clear supernatant liquid to a 150 ml beaker containing 75 ml of 95 per cent ethanol.
- ☆ Mix well and let it stand for 1 h.
- ☆ Filter through a tarred Gooch crucible.
- ☆ Wash with two 25 ml portions of 95 per cent ethanol
- ☆ Dry at 75°C for 30 min and weigh.

Observations

Sample No.	*Weight of*			*Cereal (Per cent)*
	Sample Taken W	*Gooch Crucible B*	*Gooch Crucible + Starch A*	
1.				
2.				
3.				

Calculations

$$\text{Cereal (per cent)} = \frac{(A - B)\,(1.45)\,(100)}{W/2}$$

where,

A = Weight of Gooch crucible and starch.

B = Weight of Gooch crucible

W = Weight of sample.

1.45 = Factor for converting from starch to cereal, assuming that cereals contain an average starch content of 69 per cent.

8. Estimation of Lactic Acid in Meat and Meat Products

The gram percent lactic acid can be determined for the meat samples, by a colorimetric method suggested by Barker and Summerson (1941) with slight modification.

Principle

The glucose and other interfering material of filtrate is removed with the treatment of copper sulphate and calcium hydroxide. An aliquot of the resulting solution is heated with conc. H_2SO_4 to convert lactic acid to acetaldehyde, which is then determined by reaction of p- hydroxydiphenyl in the presence of copper ions.

Reagents and Materials

1. Preparation of Dye Solution

One and half grams of p-hydroxydiphenyl is dissolved in ten ml of 5 percent NaOH and a little water by warming and stirring solution was diluted to 100 ml with distilled water. This reagent is stored in brown bottle and was kept in refrigerator.

2. Stock Solution of Lactic Acid

This was prepared from anhydrous Lithium lactate, which is anhydrous. 0.213 gram of pure dry lithium lactate is dissolved in about 50 ml of distilled water in a 100ml volumetric flask. About one ml of concentrated H_2SO_4 was added. Dilution is made to 100 ml with distilled water and it was mixed. This solution contained two milligrams of lactic acid in one ml.

3. Working Standard

One ml of stock solution of lactic acid is made to 50 ml with distilled water freshly. It contained 0.04mg/ml lactic acid.

- *20 percent Copper sulphate solution*: One hundred gram of $CuSO_4.5H_2O$ is dissolved in about 250 ml of distilled water with the aid of heat. It is then cooled and diluted to 500 ml with distilled water.
- *4 percent Copper sulphate solution*: One volume of 20 per cent copper sulphate solution is diluted to five volume distilled water and mixed.

Procedure

- ☆ To extract lactic acid, take one gm of minced meat sample was suspended in ten ml of distilled water and heated at 60°C in a water bath for 15 min. with constant stirring.
- ☆ It was then cooled in cold water, centrifuged and filtered.
- ☆ One ml of filtrate was taken in a test tube and diluted with distilled water in 1:3 ratio.
- ☆ To 2 ml of this solution, one ml of 20 per cent copper sulphate solution was added and volume made upto ten ml with distilled water.
- ☆ Approximately, one gm of Calcium hydroxide was then added to this solution and it was kept at room temperature for 30 min. with occasional stirring and after that it was centrifuged for 20 min. at 4000 rpm.
- ☆ For the blank, one ml of 20 per cent copper sulphate solution was added to two ml of distilled water. Rest of the procedure was the same as described above.
- ☆ Extracted lactic acid was estimated calorimetrically. To one ml of supernatant fluid from each of the sample as well as from the blank, 0.05 ml (one drop) of 4 per cent copper sulphate solution was added and cooled in a ice-bath.
- ☆ To the cooled solution, six ml concentrated H_2SO_4 was added while stirring constantly, and then kept in boiling water bath for five min.
- ☆ It was then cooled to 20°C. To the cooled mixture, 0.1 ml (two drops) of 1.5 percent solution of Parahydroxy diphenyl in 0.5 per cent solution of NaOH was added and stirred.
- ☆ The stirred mixture was kept in water bath for 30 min. at 30°C with occasional stirring followed by heating in boiling water bath for 90 sec.
- ☆ Solutions were cooled to 20°C in cold water and optical density (O.D.) was measured at 560 nm in a spectrophotometer.

Colorimetric estimation of lactic acid as in above was also done in the standard lactic acid solution, which was prepared freshly from lithium lactate. Here one, two, three, four and five ml of working standard of lactic acid was used which had 0.04 mg, 0.08 mg, 0.12 mg, 0.016 mg and 0.20mg of lactic acid, respectively. One ml of 20 percent copper sulphate solution was added to 1, 2, 3, 4 and 5 ml of working standard of lactic acid and made to ten ml with the help of distilled water. Rest of the procedure was the same as described above. Concentration revealing values nearer to the sample was used as standard.

Lactic acid in gram per 100 grams of meat was calculated as follows:

$$\text{Gram \% Lactic Acid} = \frac{\text{O.D. of sample} \times \text{Conc. of standard (mg)} \times \text{Dilution factor}}{\text{O.D. of standard} \times \text{wt. of sample in gms} \times 1000} \times 100$$

where, dilution factor = 10 × 4 × 5 = 200.

9. Estimation of Phytic Acid in Meat and Meat Products (Davies and Reid, 1979)

Reagents and Materials

1. 0.5 M HNO3: 15.96 ml of HNO_3 diluted with water to 500 ml.
2. Ferric Ammonium Sulphate : 216 mg Ferric ammonium Sulphate is dissolved in water to make 500 ml.
3. Ammonium thiocynate : 10 g Ammonium thiocynate is dissolved in water to make 100 ml.
4. Isoamyl alcohol.
5. Sodium phytate: 30.54g Sodium phytate (Assay 97 per cent) is dissolved in 0.5 M HNO3 to make 100 ml with 0.5 M HNO3 (0.2 mg/ml).
6. Shaker
7. Whatman Filter paper No.1

Procedure

Extraction

- ☆ One g well ground sample is extracted with 25 ml 0.5 M HNO_3 for 3 h with continuous shaking in shaker.
- ☆ Then it is filtered through Whatman No. 1 filter paper.

Estimation

- ☆ For a standard curve take concentration *i.e.* 0.4 to 1 ml of standard sodium phytate solution containing 80-200 µg Phytic acid and the volume is made to 1.4 ml with water.
- ☆ For sample take 0.5 ml of the extract in a test tube. Add 0.9 ml water.
- ☆ To all the tubes add 1 ml ferric ammonium sulphate solution and place in a boiling water bath for 20 min.
- ☆ Cool the tubes to room temperature and add 5 ml isoamyl alcohol.
- ☆ Mix the contents of the tubes by inversion and add 0.1 ml ammonium thiocynate solution.
- ☆ Shake the tubes well and centrifuge at 3000 rpm for 10 min.
- ☆ The color intensity is read at 465 nm against amyl alcohol blank exactly after 15 min of addition of ammonium thiocynate solution.

Calculations

W = Net weight (g) of sample taken

V = Volume of extract made

v = Volume of extract taken for color development

x = Concentration of Phytate contents of sample aliquot obtained from graph

$$\text{Then Phytic acid (mg/100 g)} = \frac{X \times V \times 100}{W \times v \times 1000}$$

10. Estimation of Saponins in Meat and Meat Products (Gestetnev *et al.*, 1966)

Reagents

1. Standard saponin solution: Dissolve 50 mg Saponin in acetic acid to make 100 ml.
2. 1 N H_2SO_4 in Dioxane: Water (1: 3 v/v)
3. Acetic acid (10 per cent)
4. Concentrated H_2SO_4
5. Na_2SO_4

Procedure

☆ Take 500 mg pulse flour in extraction flask and disperse in 50 ml 1 N H_2SO_4 in dioxin water (1 : 3) and hydrolyse under reflux for 8 h.

☆ Dilute the contents with addition of 50 ml water.

☆ Extract sapogenins with 25 ml diethyl ether and then with three successive portions of 15 ml diethyl ether.

☆ Combine ether extracts and wash with water and make moisture free by adding Na_2SO_4 and then dry.

☆ After drying take up the minimal amount of benzene and purify on a column of aluminium.

Column Preparation

☆ Suspend 5g freshly activated alumina (Aluminium oxide active acidic) in 100 ml benzene and 0.2 ml 10 per cent aqueous acetic acid and stir vigorously for 30 min.

☆ Pour into a column of 15 mm diameter and wash with 250 ml of benzene. Lead sapogenic extract (1.2 ml) on the column and remove various impurities by washing the column with 100 ml benzene/and elute with 100 ml 3 per cent solution of methanol in benzene.

☆ Concentrate sapogenin solution nearly to dryness and take up the residue in 10 ml acetic acid.

☆ To 2.0 ml acetic acid solution of sapogenins add 1 ml glacial acetic acid followed by 2 ml concentrated H_2SO_4.

- ☆ Cool the tubes to room temperature and read absorbance at 530 nm against a blank (3 ml glacial acetic acid and 2 ml concentrated H_2SO_4).
- ☆ Prepare standard curve by taking 0.5 ml to 3 ml standard saponin solution (0.25 kg to 1.5 mg saponin).
- ☆ 0.5 mg saponin corresponds to 0.125 OD.

11. Estimation of Tannins in Meat and Meat Products (Burn 1971)

Principle

Tannins are extracted in methanol and treated with Vanillin in acidic medium. A pink rose color is developed which is due to the formation of a complex of tannin hydroxy benzaldehyde. The intensity of the color is measured spectrophotometrically at 525 mm.

Reagents

1. *HCL-Vanillin reagent*: Just before use, mix equal volumes of 8 per cent HCl in methanol and 4 per cent Vanillin in methanol (freshly) prepared). If there appears any trace of color, reject the reagent and prepare a fresh.
2. *Standard Catechin solution*: (2 mg/ml). 100 mg catechin is dissolved in methanol and diluted to 50 ml with methanol.

Procedure

Extraction

- ☆ Take 10 g ground seeds in a 150 ml Soxhlet flask and reflux with 25 ml methanol for 3-4 hours.
- ☆ Filter the extract through Whatman No. 1 filter paper.

Determination

- ☆ For a standard curve take 0.1 ml to 0.6 ml standard catechin solution (2mg/ml) in test tubes.
- ☆ The total volume of each tube is made to 1.0 ml with methanol.
- ☆ For sample, take 1 ml of the clear filterate in a test tube.
- ☆ For blank take 1 ml methanol.

- ☆ To all the test tubes add 5 ml vallinin HCl reagent.
- ☆ After 20 min. read the color intensity at 525 nm.
- ☆ These values are expressed as catechin equivalents.

Calculations

W = Net weight (g) of sample taken

V = Volume of extract made

v = Volume of extract taken for color development

X = Concentration of tannins obtained from graph

$$\text{Tannins (mg/100g)} = \frac{X \times V}{W \times v} \times 100$$

12. Estimation of Total Polyphenolic Compounds

Principle

Folin Denis reagent produces a blue color with polyphenolic substances. Tannic acid is used as a standard and results are expressed as tannic acid equivalents.

Reagents

1. *Folin Denis Reagents*: To 750 ml of water added 100 mg Sodium tungstate ($Na_2WO_4.2H_2O$), 20 g phosphomolybdic acid and 50 ml phosphoric acid. Reflux it for 2 h, cool and make upto 1 litre with Distilled water.
2. *Saturated Sodium carbonate solution*: To 100 ml water added 45 gm anhydrous Sodium carbonate, dissolve it at 70-80° C and let it cool overnight. After crystallization, filter it through glass wool.
3. *Tannic acid standard solution*: Dissolve 100 mg tannic acid in 1 litre of water. Prepare fresh solution for each determination.
4. *Methanol-HCl*: 10 ml concentrated HCl in methyl alcohol and make volume 1 litre with the methanol (methyl alcohol)

Procedure

- ☆ Weigh 200 mg defatted material in a 250 ml round bottom flask

- Added 100 ml methanol HCl and reflux it for 2 h and allow to cool.
- Filter extract through Whatman's paper No. 40 into a 100ml volumetric flask and make up volume 100 ml after few washings with methanol- HCl.
- Take 0.2 ml of extract for estimation of polyphenols.
- To this, add methanol-HCl to make volume 7.5 ml
- Add 1 ml saturated Sodium carbonate to each
- Add 0.5 ml F.D reagent.
- Add 1 ml distilled water to make 10 ml.

Standard Curve

- It is made by taking 0, 0.2, 0.4, 0.6, 0.8 and 1.0 ml standard tannic acid.
- To this add methanol- HCl to make volume 7.5 ml. Add distilled water to make volume 10 ml.

Calculations

W = Weight of sample

100 ml (V) = Aliquot made

V_1 = Aliquot taken

C = Concentration graph

$$\text{mg Polyphenol/100g} = \frac{C \times V \times 100}{V_1 \times W}$$

Chapter 6
Biochemical Methods

1. Spectrophotometry

Absorptiometeric Analysis

A spectrophotometer is a sophisticated type of colorimeter where monochromatic light is provided by a grating or prism. The bandwidth of the light passed by filter is quite broad, so that it may be difficult to distinguish between two compounds of closely related absorption with a colorimeter. A spectrophotometer is then needed, when the two peaks can be selected on the monochromator.

Principle

It is based on the principle of light absorption. Every chemical substance has a characteristic atomic and molecular structure. Due to this, the chemical substances have specific resonance frequency. The light, according to electromagnetic theory, travels in the form of a wave.

Light wave → Wavelength
Light wave → Frequency
Light wave → Energy in photons

$$E = h\nu$$

where,

E = Energy of light

h = Plank's constant

ν = Frequency/sec

Intensity of beam of radiation is characteristic by its radiant power which is proportional to number of photons propagated per sec. Light may be monochromatic (carry one discrete wavelength) and polychromatic (carry several wavelengths).

Absorption

When light falls on a chemical compound, the emitted light has decreased amount of energy than the incident light due to transition of energy. This phenomenon of energy transfer is called absorption. Amount of radiant light absorbed at a particular wavelength is proportional to the concentration of the solution. Color of the solution is due to the selective absorption of light at certain visible wavelength.

Lambert's Law

When the ray of monochromatic light passes through an absorbing medium its intensity decreases exponentially as the length of absorbing medium increases

According to Lambert's law,

$$I = I_0 e^{-K_1 l}$$

where,

I_0 = Incident beam intensity

I = Emerged beam intensity

l = Depth of solution

K_1 = Constant

Beer's Law

When a ray of monochromatic light passes through an absorbing medium, its intensity decreases exponentially as the concentration of the absorbing medium increases

According to Beer's law,

$$I = I_0 e^{-K_2 C}$$

where,

C = Concentration of absorbing solution

Beer-Lambert Law

When the ray of monochromatic light of initial intensity (I_0) passes through a solution in a transparent vessel, some of the light is absorbed so that the intensity of the transmitted light (I) is less than (I_0). There is some loss of light intensity from scattering by particles in the solution and reflection at the interfaces, but mainly by the absorption by the solution. The relationship between I_0 depends on the path length (l) and concentration (c) of the absorbing medium. These factors are related in Beer-Lambert law

According to Beer-Lambert law

$$I = I_0 e^{-K_3 Cl}$$

Transmittance (T)

It is the ratio of intensities and expressed as

$$T = I/I_0 = e^{-K_3 Cl}$$

Taking logarithms,

$$\mathrm{Log}_e I_0/I = K_3 Cl$$

$$\mathrm{Log}_{10} I_0/I = 2.303\ K_3 Cl$$

$$\mathrm{Log}_{10} I_0/I = KCl$$

The expression $\mathrm{Log}_{10} I_0/I$ is known as extinction E or absorbance A. The extinction is sometimes referred to as optical density (OD).

Therefore E = KCl

If Beer-Lambert's law is obeyed and l is kept constant say 1 cm, then a plot of extinction against concentration gives a straight line passing through the origin, where as plot of per cent transmittance against concentration gives negative exponential curve.

Visible range = 360-900nm

Ultraviolet = 200-360nm

Color	*Maximum Transmission*
Violet	380-435nm
Greenish blue	435-470nm
Bluish green	470-500nm
Green	500-560nm
Yellowish green	560-590nm
Yellow	590-650nm
Orange	650-700nm
Red	More than 780nm
Near infrared	>800nm

Limitations of Beer-Lambert Law

- ☆ Sometimes a non-linear plot is obtained by OD against concentration of the absorbing solution. This is probably due to one or other of the following conditions being not fulfilled:
- ☆ Light must of narrow wavelength and monochromatic
- ☆ The wavelength of the light used should be at absorption maxima (lmax) of the solution. This also gives the greatest sensitivity.
- ☆ There must be no turbidity in the solution; the solution should not be much concentrated.
- ☆ There should be no ionization, no dissociation and no association with concentration and time of solute.

2. Colorimetery

The measurement of concentration of colored substances usually in the solution for the basis of colorimetric analysis. Colorimetery is based upon matching of a colored solution representing an unknown concentration of the substance with standard curve made out of standard solutions containing the substance in unknown concentrations. The substance must therefore be colored by itself or capable of undergoing reactions leading to the production of a colored complex in solution. The color intensity is dependent upon concentration of the compound. A colorimetric procedures thus involves three operation:

- ✰ The preparation of a colored solution
- ✰ Preparing standard colored solution
- ✰ Drawing of a standard curve and matching of the concentration of the unknown sample

A diagram of the basic arrangement of a typical colorimeter shows that white light from a tungsten lamp passes through a slit, then a condenser lens, to give a parallel beam, which falls on the solution under investigation contained in an absorption cell or cuvette. The cell is made of glass with the sides facing the beam cut parallel to each other. In most cases the cells are 1 cm^2 and will hold 3 ml of liquid comfortably. Beyond the absorption cell is the filter, which is selected to allow maximum transmission of the color absorbed. If a blue solution is under examination, then red is absorbed and a red filter is selected. The color of the filter is therefore complimentary to the color of the solution under investigation. In some instruments the filter is located before the absorption cell. The filter gives narrow transmission bands and therefore, approximate to, monochromatic light. The light is then falls on to a photocell which generates an electrical current in direct proportion to the intensity of light falling on it. This small electrical signal is increased in strength by the amplifier. The amplified signal passes to a galvanometer, which is calibrated with a logarithmic scale, so as to give absorbance readings directly.

Relationship Between the Colors of the Solution Examined and Filter Chosen for Colorimetric Analysis

Color of the Solution	*Filter*
Red-Orange	Blue-blue green
Blue	Red
Green	Red
Purple	Green
Yellow	Violet

The blank solution is first put in the colorimeter and the galvanometer adjusted to zero extinction, this is followed by the test solution and the extinction is read off directly.

(*i*) Estimation of Amino Acids (Paper Chromatography)

Principle

Free amino acids in a protein-free filtrate are separated by paper chromatography. The amino acid ninhydrin color complex is converted to a deeply colored copper complex and then evaluated densitometrically.

Reagents

1. Ninhydrin reagent 0.5 per cent in pure acetone
2. N–butanol : acetic acid: water (1.2 : 3 : 5).
3. Copper nitrate solution : 1 ml of standard copper nitrate and 0.2 ml of 10 per cent HNO_3 made upto 100 ml with pure acetone.
4. Standard amino acid mixture
5. Amino acid mixture.

Valine, Alanine, Lysine, Leucine, Tyrosine, Glycine, Serine, Histidine, Iso-leucine, Tryptophane. Phenylalanine, Arginine, Methionine, Aspartic acid, Cystine, Glutamic acid, Glutamine.

Take 20 mg of each of the amino acids dissolve in H_2O, add few drops of 1N HCl to 100 ml before spotting.

Procedure

Preparation of Sample

(a) Food Stuffs

☆ Take 1g powdered sample and extract in 100 ml of 70 per cent alcohol by occasional stirring for overnight.

☆ Filter the extract through Whatman No. 1 and evaporate the filtrate to dryness on a steam bath. (If Glutamic acid is to be determined, dry the extracted filtrate at room temperature using vacuum oven).

☆ The residue is then taken up in 10 ml of 10 per cent Iso-propanol and filter.

☆ The clear extract is used for spotting. The extract is stable at 4°C. Twenty ml of this extract is used for spotting (For two way run, use 100 ml of extract).

(b) Plasma/Serum

- Take 1 ml of serum or plasma. Add 9 ml of 80 per cent of ethanol.
- Shake well and centrifuge.
- The supernatant solution is evaporated to dryness and the residue is taken up with 1 ml of H_2O or isopropanol.
- Use 100 µl for spotting

Spotting

- Take a Whatman No. 1 paper sheet.
- From one side, a line of application is marked with ordinary pencil leaving a space of 1.5.
- The points of application of sample and mixture of amino acids are marked on the line.
- Take 20 µl of iso-propanal extract of the sample and spot on the filter paper sheet with occasional drying with hot air.
- Similarly spot for the amino acid mixture.
- Irrigate the spotted paper with n-butanol-acetic acid water solvent system for 17 h by descending technique in a chromatography chamber.
- Then dry the paper in oven and spray with 0.25 per cent ninhydrin solution in acetone.
- The sprayed paper is air dried for 10 min and heated at 50°C for 1 h and left overnight in dark.
- Drying at low temperature gives a clean background and spots develop evenly.
- The paper is then sprayed with copper nitrate solution. The copper complex color is more intense and the chromatogram can be preserved for a long time.

Identification of Spots

- The spots are identified on the basis of their mobility in a solvent system.
- This mobility is expressed by RF values.
- It is the ratio of the distance, the compound moves along the paper to the distance covered by the solvent.

$$RF = \frac{\text{Distance covered by the sample}}{\text{Distance covered by the solvent}}$$

The RF values of the compounds, which correspond to the RF values of those of the mixture spots of the standard amino acid, spots, help to locate the particular amino acids of the sample.

Measurement of the Color Intensity of the Spots

- ☆ The area of each spots is cut out and dissolved in water and volume is made to 10 ml.
- ☆ The color intensity is measured at 510 nm.
- ☆ By comparing the intensity of unknown spot with that of the corresponding standard spot, the concentration can be calculated.

W = Net weight (g) of sample
V = Volume of extract made
X = Concentration of standard Amino acid

The concentration of amino acid (mg/100g) =

$$\frac{\text{O.D. of unknown} \times X \times V \times 100}{\text{O. D. of standard} \times W \times 1000}$$

3. Separation Techniques

(*i*) Exclusion Chromatography (Gel Filtration)

Principle

The term gel filtration is used to describe the separation of molecules of varying molecular size utilizing these gel materials. The general principle of exclusion chromatography is quite simple. A column of gel particles is in equilibrium with a suitable solvent for the molecules to be separated. Large molecules, which are completely excluded from the pores pass through the interstitial spaces, while smaller molecules are distributed between the solvent inside and outside the molecular seive and then pass through the column at a slower rate. The solvent absorbed by a swollen gel is available to a solute to an extent, which is dependent upon the porosity of the gel particle and the size of the solute molecules. Thus the distribution of a solute in a column of a swollen gel is determined solely by the total volume of solvent, both inside and outside of the gel particles, which is available to it.

$$Vt = Vi + Vo$$

when,

Vt = Total bed volume

Vi = Volume of hydrated beads

Vo = Void Volume *i.e.* liquid between the gel beads.

Reagents and Materials

1. Sephadex gel column
2. Hemoglobin (red)-large molecule having mol wt. 68,000
3. Methylene blue- small molecule molecular weight 200

Method

The sephadex gels are obtained by cross-linking the polysaccharide dextran with epichlorhydrin. In this way the water-soluble dextran is made water insoluble, but it retains its hydrophilic character and swells rapidly in aqueous media forming gel particles suitable for gel filtration. The gel particles act as molecular sieve.

2.5 g of sephadex G-25 soaked in buffer solution of 0.05 M Tris HCl at pH 7.4 overnight and the slurry is made. The slurry poured in the column using a glass rod to prevent entrapping of air. The column was packed with material by filling it one third full with solvent and slowly adding the slurry of the material in the solvent. This is carefully poured down with a glass rod, to stop air bubbles being trapped in the column. The suspension was allowed to settle and excess solvent run off. After thoroughly washing the column with solvent, the level of the liquid is allowed to fall to just above the surface of the material. Finally a filter paper disc is placed on the top of the adsorbent to avoid disturbing the surface when the sample is applied.

Application of the Sample

200 µl of Hemoglobin and 100 µl of Methylene blue solution were mixed. The solution mixture was evenly spread across the

surface of the column bed. The tap is opened until the top of the column is just below the level of the meniscus. The solvent reservoir is connected and a constant head of liquid maintained at the top of the column from a pressure reservoir.

Elution

The next stage, the materials from the column in order by eluting with an appropriate solvent. Here Tris buffer was used. For better resolution, elution development is preferred. The solvent interacts with the column more weakly than the solute molecules and overrides the bound molecules gradually eluting them from the column and different molecules are removed from the column by changing the pH, ionic strength or polarity of the solvent. This can be done by simply changing the eluting solvent when one compound emerges from the column and this is known as stepwise elution

The mixture of the smaller and larger molecules passes through the gel column. The smaller molecules can enter the water that is present as a stationary phase inside the gel particles, whereas the larger molecules pass through or extruded. The smaller molecules constantly move inside and outside the gel particles and hence partition themselves between the two volumes of water that are present inside and outside the gel network.

Collection and Analysis of Fractions

The effluent from the column was collected in 2 ml each in a series of test tubes manually. Each fraction was then analyzed for the presences of the compounds being examined. The absorbance or OD was recorded at 405 nm and 660 nm, respectively for haemoglobin and methylene blue. An elution profile was prepared for the quantity of the compound eluted against the effluent volume.

(*ii*) Subcellular Fractionation and Differential Centrifugation

Principle

This method is based upon the differences in the sedimentation rate of particles of different size and density. Centrifugations will initially sediment the largest particles. For particles of the same mass but different density, the ones with highest density (*e.g.* peroximes) will sediment at faster rate than the less dense particles (*e.g.* plasma membranes) particles having similar banding densities *i.e.* most of the Subcellular organelles (where p =1.1 to 1.3 g cm^{-3} in sucrose solution) can usually be efficiently separated from one another by differential centrifugation or the rate zonal method.

In differential centrifugation, the material to be separated (*e.g.* tissue homogenate) is centrifugally divided into a number of fractions by increasing (step-wise) the applied centrifugal force.

Extraction

A common solution for extraction of organelles consists of sucrose 0.25 mol/l (iso-osmotic) adjusted to pH 7.4 by Tris (Tris [hydroxy methyl] amino methane)-hydrochloric acid buffer 0.005 mol/l containing K^+ and Mg^{2+}ions at neat physiologic concentration (STKM solution). Tissue used was goat liver. Extraction is carried over in cold room at 4°C, because proteolytic enzymes are active at room temperature

Homogenization

25 per cent liver homogenate was prepared by taking 1g liver and 3 ml sucrose-Tris buffer using a tissue homogenizer resulting a suspension containing many organelles in intact condition. This suspension is called homogenate.

Centrifugation

Initially all particles of the homogenate are homogeneously distributed throughout the centrifuge tube. During centrifugation, particles move down the centrifuge tube at their respective sedimentation rates and start to form a pellet on the bottom of the tube. In the time required for the complete sedimentation of heavier particles, some of the lighter and medium sized particles originally suspended near the bottom of the tube, will also sediment and thus contaminate the fraction. Pure preparation of the pellet of the heaviest particles cannot therefore be obtained in one centrifugation step. The separation achieved by differential centrifugation may be improved by repeated (2 or 3 times) resuspension of the pellet in homogenization medium and recentrifugation under the same conditions as in the original pelleting. Further centrifugation of the supernatant in gradually increasing centrifugal fields results in the sedimentation of the intermediate and finally the smallest and least dense particles. A scheme for the fractionation of the goat liver homogenate into various Subcellular fractions is also given. In spite of its inherent limitations, differential centrifugation is probably the most commonly used method for the isolation of cell organelles from homogenized tissue.

(*iii*) Sodium Dodecyl Sulphate Polyacrylamide Gel Electrophoresis (SDS-PAGE)

SDS-PAGE Method as Described by Laemmli (1970)

Samples were prepared by heating them for 5 minutes in a boiling water bath after addition of an equal volume of 2 x Laemmli dissociating buffer (0.125 M Tris-HCl, pH 6.8, 4 per cent SDS, 20 per cent glycerol, 10 per cent 2-mercaptoethanol). Separating gel contained 10 per cent acrylamide in 0.125 M Tris-HCl pH 6.8. Both contained SDS at 0.1 per cent level. The electrode buffer consisted of 0.025 M Tris, 0.193 M glycine solution containing 0.1 per cent SDS Electrophorsis was carried out at constant voltage of 60 v over night by which time the bromophenol blue marker dye reached the bottom of the gel.

Fixing and staining of the gel was simultaneously accomplished using 0.125 per cent coomassie Brilliant Blue R-250 solution in methanol, acetic acid mixture. Destaining was conducted in the same solution without the dye until bands were clearly visible. Purified rat-tail collagen and rainbow™ protein molecular weight marker were run simultaneously.

Sodium Dodecyl Sulphate Polyacryiamide Gel Electrophoresis (SDS-PAGE)

SDS-PAGE of meat extractable proteins was carried out as per the method of Laemmli (1970) with slight modification using 1 mm slab gel in Genei electrophoresis equipment (Bangalore) and power supply apparatus (Pharmacia).

Preparation of Sample

Meat samples were taken for SDS-PAGE. Meat proteins were extracted from meat sample following the procedure of Ishikawa *et al*. (1987). 500 mg of meat sample was homogenized in 5 ml of ice-cold 0.1M Ammonium bicarbonate solution (pH 8.6) by 3 or 4 strokes using tissue homogenizer. The homogenate was centrifuged using a micro-centrifuge at 3000 rpm for 15 min. The supernatant containing extractable protein was collected and kept at –20°C for use in SDS-PAGE.

SDS-PAGE Reagents

1. *4 × Lower buffer (1.5M Tris/HCl, pH 8.8)*: 18.17g Tris base was dissolved in 90 ml distilled water and adjusted pH with concentrated HC1 to 8.8. Then 0.4g SDS was dissolved in it. Final volume was made to 100 ml with distilled water.
2. *Acrylamide 30 per cent Bisacrylamide 0.8 per cent*: 30g Acrylamide and 0.8g N.N′-methylene bisacrylamide were taken in a 150 ml conical flask and 60 ml distilled water was added to it. Stirring was continued till the chemicals dissolved and volume was made to 100 ml. 2-teaspoonful activated charcoal was added, stirred for 5 min and filtered to get clear solution.
3. *Sodium dodecyl Sulphate (SDS) 10 per cent*: Dissolved 5g SDS in 45 ml distilled water and volume was made to 50 ml.
4. *Upper buffer (0.5 M Tris/HCl, pH 6.8)*: 6.05g Tris base was dissolved in 80 ml distilled water and pH was adjusted to 6.8 with concentrated HC1. Volume was made to 100 ml.
5. *Ammonium persulphate (APS) 10 per cent*: It was prepared freshly by dissolving 100 mg APS in 1 ml distilled water.
6. *2 × Sample buffer*: 2 ml glycerol, 1 ml 2-mercaptoethanol, 4.5 ml 10 per cent SDS, 1.7 ml upper buffer and 0.2 ml 0.1 per cent bromophenol blue were taken and 0.6 ml distilled water was added to them. The solution was mixed properly and stored at 20°C.
7. *Electrode buffer (pH 8.3)*: 6g Tris base and 28.8g glycine was dissolved in 1500 ml distilled water. The pH was checked, if not 8.3, adjusted with IN HCl. 2g SDS (or 20 ml 10 per cent SDS) was added to it and the final volume was made to 2 lt.
8. *SDS-PAGE gel stain*: 1.25g Coomassie blue (R 250), 500 ml methanol, 100 ml acetic acid and 400 ml distilled water were mixed, stirred for 30 min and filtered to get 1 litre solution.

9. *SDS-PAGE gel destain*: 300 ml methanol, 70 ml glacial acetic acid and 630 ml distilled water were mixed properly to get 1 lt. solution.
10. *Separating gel (10 per cent)*: For 10 ml Acrylamide/bisacrylamide solution, 7.5 ml deaerated 4 x lower buffers and 12.4 ml distilled water were mixed gently in a beaker. Then freshly prepared 74 µl of 10 per cent ammonium persulphate and 10 µl N.N.N^1, N^1 tetramethylenediamine (TEMED) were added immediately and mixed.
11. *Stacker gel (5 per cent)*: 1.3 ml Acrylamide/bisacrylamide solution, 2.5 ml upper buffer, 0.1 ml 10 per cent SDS and 6.1 ml distilled water were mixed gently in a beaker. To this, 50 µl ammonium persulphate and 10µl TEMED were added and mixed.

Procedure

- ✰ *Casting of gel*: 2 per cent–melted agar was poured from side through spacers to seal the bottom and side of the glass plates. Then 10 per cent separating gel was poured carefully into the gel casting space between the glass plates until about 75 per cent of the space was filled. Water saturated *n*-butanol was layered over the gel and after polymerization of the separating gel; n-butanol was drained off by tilting the gel cast assembly. The gel upper surface was then washed with distilled water to remove n-butanol, if any. 5 per cent stacker gel solution was layered over the separating gel after washing the upper surface by the same gel solution. Slot forming comb was carefully inserted into the top of the gel casting area until both ends of the comb were stopped at tops of the side spacers. Water saturated n-butanol was overlayered. After polymerization of stacker gel, the comb was removed slowly and carefully.
- ✰ *Running of gel*: Protein samples and molecular weight marker (MW-SDS-200 kit) were diluted with equal volume "of 2′ sample buffer and boiled in hot water bath for 5 min. First molecular weight marker protein of about 10 µl was applied. Then samples were applied to each slot so that amount of protein was 200 µg in each case. The upper

chamber of the electrophoresis apparatus was filled with about 600 ml electrode buffer and about 1400 ml buffer was poured into the lower chamber. Electrophoresis was performed at a constant voltage mode of 50V/slab at 13mA for first 1 h and increased to 100 V/slab subsequently and maintained at same voltage for about 5 to 6 h or till the tracking dye reached the lower end of the gel. Electrophoresis was carried out at room temperature.

- ☆ *Removal of gel*: Power supply was turned off and the electrophoresis unit was disassembled. Removed clamps from each gel sandwich. The spacer bars were slided out and the glass plates were separated with syringe handle. Nicked the gel at appropriate place (lower corner opposite molecular weight marker, dye front edge of molecular weight strip). Then stacking gel was cut off.
- ☆ *Fixing, staining and destaining of gels*: The gels after electrophoresis were fixed for overnight with fixing solution (10 per cent acetic acid and 50 per cent methanol in distilled water) and stained with the gel stain solution for 4 to 5 h. The gels were then destained with several changes of gel destaing solution. After thorough destaing, the gels were stored in 7 per cent acetic acid till photographed.

(*iv*) Isolation of DNA from Buffalo Liver

Principle

DNA is very easily damaged by the shear force. Even a rapid storing of solute can break high molecules of DNA into smaller fragments. So extraction procedure is gentle method of cell disruption. The second damaging factor is nuclease *i.e.* DNase which are found in most of the cells, and is also present in dust and fingers which would contaminate glasswares during oxidation of nucleic acid. Cell disruption should be done at –4°C. When the DNA come to solution RNase treatment is given to remove RNA. Protein is removed by using chloroform-Phenol mixture that denatures the protein but not DNA.

Reagents and Materials

1. Sodium dodecyl Sulphate

2. Tris buffer
3. Phenol
4. Chloroform
5. Ethanol
6. Sodium chloride
7. Refrigerated centrifuge
8. Hot water bath
9. Deep freezer

Procedure

- 1 ml of nuclear suspension was taken.
- 1 ml of Tris buffer pH 7.5 and 1 per cent Sodium dodecyl Sulphate (SDS) of total concentration here 0.2 g was added to it.
- The solution is then kept at 65°C for 30 min.
- Added 2 ml of Tris buffer to make total volume to 4 ml.
- Again 4 ml super saturated phenol- Chloroform mix was added to it and centrifuged briskly at 3000rpm for 5 min.
- The upper aqueous layer contained the DNA.
- To the aqueous layer 1/10 volume NaCl 1 M solution (here 0.2 ml) mixed with ethanol 2.5 volume was added.
- Then it was kept at –20°C, left overnight.
- The buffer was added and centrifuged.
- The precipitate was taken after throwing the supernatant, and it was dissolved with100 µl-distilled water and volume was made upto 3ml by distilled water.
- Then OD of the solution was recorded at 260nm.

Observations

OD of the sample =

(*v*) Estimation of Deoxy Ribonucleic Acid (DNA) by Diphenylamine Method

Principle

DNA when heated with perchloric acid releases deoxy-ribose, which reacts with hot TCA giving rise to ω-*hydroxy-levuilaldehyde,* which reacts with diphenylamine giving rise to blue color *i.e.*, read at 565 nm in a colorimeter or spectrophotometer.

Reagents

1. Standard DNA solution (1mg/ml) Standard DNA solution is prepared in 1N perchloric acid heating at 70°C for 15 min
2. Diphenylamine reagent

Composition

Diphenylamine: 1.5g

Concentration H_2SO_4: 1.5 ml

Glacial acetic acid: 100 ml

Procedure

- ☆ The standard DNA solution (1mg/ml) was taken in different concentration of 50µg, 100µg, 150µg, 200µg, 250µg and 300µg in the test tubes numbered 1, 2, and 3,4,5,6 respectively. For these DNA standard solutions of 0.05 ml. 0.1 ml, 0.15 ml, 0.2 ml, 0.25 ml and 0.3ml were taken in the above order.
- ☆ Test tube No. 7 was used for blank reagent.
- ☆ In test tube No. 8, unknown sample 0.1 ml was taken.
- ☆ Using distilled water, the volume of each tube was made upto exactly 2 ml.
- ☆ 3 ml of Diphenylamine reagent was added to each test tube and mixed properly. This reagent was prepared fresh before use.
- ☆ All the test tubes then kept in boiling water bath for about 15 min and thereafter the test tubes were cooled. The blue color thus developed was read at 565 nm.

(*vi*) Estimation of Ribonucleic Acid by Orcinol Method

Principle

The ribose sugar present in the RNA is converted into furfural due to the action of concentrated HCl and this furfural react with Orcinol giving blue-green color whose absorbance (O.D.) is read at 675 nm to find out the RNA content

Reagents

1. 0.1 per cent ferric chloride ($FeCl_3$) solution in concentrated HCl.
2. 0.2 per cent Orcinol solution in absolute alcohol
3. Standard RNA solution : Crystalline RNA is dissolved in 0.1 N NaCl to get a concentration of 5 mg/ml. If cloudy, a few drops of 0.1 N NaOH is added, filtered and stored in cold. The standard RNA solution supplied in this experiment had concentration of 1mg/ml.

Procedure

- ☆ For standard solutions of RNA, five test tubes were numbered 1,2,3,4 and 5.
- ☆ The supplied stand RNA solution (1mg/ml) was further diluted 10 times with distilled water so that it contained 0.1mg/ml or 100 µg/ml.
- ☆ Using this standard solutions, RNA standard concentration of 25 µg, 50 µg, 100 µg, 200µg and 300µg were taken in test tubes 1,2,3,4 and 5 respectively *i.e.*0.25ml, 0.5ml, 1.0ml, 2ml and 3ml respectively of standard RNA solution were taken in the above test tubes in serial order.
- ☆ In test tube 6 blank reagents was taken and in the test tube no 7, 0.1 ml of the unknown sample was taken.
- ☆ The volume of each test tube was made upto 3 ml using distilled water. 3ml of $FeCl_3$ solution and 0.3 ml Orcinol reagent was added to each test tube and the contents were mixed thoroughly.
- ☆ Then all the test tubes were kept in the boiling water bath for 20 min.
- ☆ After that, the test tubes were cooled and O.D. or absorbance of solution in each test tube was recorded at 675nm. Standard curve was plotted taking concentration of RNA (µg) in X-axis and O.D. in Y-axis of a graph paper.
- ☆ Using the standard curve, the concentration of RNA in the unknown sample was determined.

(*vii*) Estimation of R-values (Inosine/Adenosine Ratio)

The estimation of R-value is carried out by following the method of Honikal and Fischer(1977) Bernthal *et al.* (1991)

Reagents and Materials

1. Perchloric acid
2. Potassium phosphate buffer
3. Homogenizer
4. Filter paper
5. Spectrophotometer

Procedure

- ✰ Homogenize 2 g of meat with 10 ml of 1.0 M perchloric acid.
- ✰ The homogenate was filtered and 0.1 ml of the filtrate was added to 4.9 ml of 0.1 M potassium phosphate buffer pH 7.0.
- ✰ Absorbance was measured at 250 nm and 260 nm using the phosphate buffer as reference.
- ✰ The R-value was defined as the ratio of absorbance at 250nm over the absorbance at 260 nm (Honikel and Fischer, 1977), and estimates the inosine/adenosine ratio in muscle.
- ✰ Duplicate determinations were made in each treatment.

(*viii*) Isolation of Proteins from Biological Source *i.e.* Plasma or Tissue Homogenate

Acid Precipitation

Principle

Due to acidic pH of the solution, the proteins, which have +ve charge, are neutralized by bonding to a large negative ion, which results into a precipitate. In acid precipitation method, the protein is denatured, so further characterization of protein is not possible in this method.

Procedure

- ✰ Homogenate containing soluble proteins of buffalo pituitary was used.

☆ 1 ml of homogenate was taken in a test tube and 2ml of Trichloro acetic acids (TCA) 16 per cent solution or 1 M solution was added to it and mixed properly.

Results

☆ Heavy precipitation was obtained.

Heavy Metal Precipitation

Principle

Proteins contain acidic amino acids having negatively charged polar R-groups such as glutamic acid and aspartic acid. Coo^- combines with positively charged heavy metal ions *viz.* Cu^{++} or Hg^{++} at pH > Pi (isoelectric pH of protein) causing precipitation of protein. Most of the proteins are slightly acidic at isoelectric point. So heavy metal precipitation is done at neutral pH or neutral solution.

Procedure

☆ 1ml of the above tissue homogenate (buffalo pituitary) was taken in a test tube

☆ 1 ml of 2 per cent $CuSO_4$ solution was added to it and mixed properly.

☆ Then it was noticed whether there was precipitation or not.

☆ Presence of precipitate means positive results

Inference

Protein is denatured by this method. For blood deproteinization, phosphotungstic acid or $ZnCl_2$ can be used.

Salt Precipitation

Principle

At high ionic strength of Ammonium Sulphate $(NH_4)_2SO_4$ solution, protein may be almost completely precipitated from the solution, an effect called salting out. The cause of salting out phenomenon may be the high concentration of salt remove the water of hydration from the protein molecules and then reducing the solubility, but other factors are also involved. Proteins precipitated by salting out retain their native conformation and can be dissolved for further studies.

Procedure

- ☆ 0.5 ml of homogenate was taken in a test tube
- ☆ 0.5 ml of saturated solution of Ammonium Sulphate $(NH_4)_2SO_4$ was added and mixed properly.
- ☆ It was observed whether there was precipitate or not.

Observations and Results

Precipitate was found.

Solvent Precipitation

Principle

Addition of water miscible neutral organic solvents *e.g.* ethanol or acetone, decreases the solubility of protein to such an extent that they precipitate out of the solution. Protein solubility at a fixed pH and ionic strength is a function of the dielectric constant of the medium. Ethanol has a lower dielectric constant than water. Its addition to an aqueous solution of protein increases the attraction force between opposite charges and decreases the degree of ionization of R-group of proteins. So protein molecules aggregate and precipitate.

Procedure

- ☆ 400 ml (0.4 ml) of the tissue homogenate was taken in a test tube and 1600 ml or 1.6 ml chilled ethanol was added to it and mixed properly.
- ☆ Then it was notice whether there is precipitate or not.

Observations

Precipitate was present. 0.5 ml of homogenate was mixed to 0.5 ml of 2 M HCl solution. It was observed that there was no precipitate.

Due to acidic pH all the protein molecules have same +ve sign, so they repel each other.

4. Lipid Composition

(i) Lipid Extraction

Many solvents or solvent combinations can be used to extract lipids from food products, but care must be taken to ensure that lipolytic and other enzymes are deactivated and that the recovery is complete, In addition, precautions must also be taken to avoid or at least minimize auto or enzymatic oxidation of polyunsaturated fatty acids fatty acids

The extraction procedure will depend on the nature of the food matrix as well as the lipid components. Neutral lipids (triacylglycerols) can be extracted by nonpolar solvents such as petroleum ether, hexane, or supercritical carbon dioxide. For the extraction of polar lipids (phospho and glycolipids) polar solvents must be used for quantitative determination, sometimes preceded by a chemical hydrolysis to break bonds formed between lipids and proteins.

Practical Protocols

Although chloroform/methanol-based solvent systems for the extraction of lipids represent major environmental hazards, they are still the rule, not the exception, in lipid research. The most common are the "Folch" and the Bligh and Dyer methods.

1. "Folch" Procedure

For the "Folch" procedure, it is essential that the ratio of chloroform, methanol and saline solution in the final mixture be close to 8:4:3. On the other hand, some variation in the approach to attaining the optimum concentrations may be possible. Detailed protocols may be found in Folch *et al.* (1957), Sempore and Bezard (1977). Briefly, the food sample is homogenized with chloroform-methanol (2: 1) to a final dilution of 20 ml per gram of sample. Then, 0.2 times the volume of 0.37 M potassium chloride is added, the mixture is homogenized and centrifuged to separate it into two phases, and the upper phase is removed by siphoning. The upper phase is replaced with chloroform-methanol-water (5:48:47), followed by homogenization and centrifugation. The upper phase is eliminated again and, if necessary, a second washing is applied. The lower phase is filtered and the residue rinsed with chloroform and chloroform-methanol (1:1). Organic extracts are joined, and the solvent is evaporated under vacuum below 40°C. The lipids are finally stored at –20°C in a small volume of chloroform with an antioxidant added.

2. Bligh and Dyer Method

This procedure was developed as an economical method for extracting the lipids from large volumes of wet tissue, from frozen fish specifically, with the minimum volume of solvent. The method also uses chloroform-methanol for the extraction; but the quantities are such that when mixed with water in the tissue, a single-phase solution is formed. Alterations to the procedure may be made but the ratios of chloroform-methanol-water before and after dilution must be 1:2:0.8 and 2:2:1.8, respectively. For applying the Bligh and Dyer method (1959), it is necessary to know the water content of the food. Briefly, for a product containing about 80 per cent water and 1 per cent lipid, the food sample would be homogenized in three volumes (in milliliters) of chloroform methanol (1:2) per gram of sample. Then, the homogenate is diluted with one volume of chloroform and one volume of distilled water (the homogenate is blended after each addition), and filtered filtrate is allowed to settle for a few minutes to complete the separation and clarification of the phases, and tile upper phase is eliminated. The lower phase, containing the lipid, is evaporated under vacuum below 40°C. The lipids are finally stored at –20°C in a small volume of chloroform with an antioxidant added.

General Precautions for Preventing Contamination and Chemical Alteration of Lipid Samples during Analysis

To Avoid Contamination

1. Distill all solvents before use
2. Rinse all glassware with chloroform-methanol (2:1, v/v) immediately before use
3. Avoid sample contact with any rubber or plastic except Teflon
4. Store samples and solvents in glass containers with glass or Teflon-lined caps

To Prevent Oxidation of Polyunsaturated Acids

1. Perform all laboratory manipulations under nitrogen
2. Bubble nitrogen or helium line through solvents to purge dissolved oxygen
3. Evaporate solvents below 40°C under an inert atmosphere

4. Add 0.005 per cent antioxidant (w/v) to solutions of stored samples*.
5. Store samples or lipid extracts under nitrogen at –20°C

To Avoid Chemical Alteration

1. Extract lipids immediately after food sampling
2. Avoid prolonged sample contact with any alcohol, particularly under acid or alkaline conditions
3. Remove the traces of water by co-distillation with pure ethanol
4. Do not leave lipids in the dry state. Cover with an inert nonalcoholic solvent such as hexane

* Care should be taken to select an antioxidant that does not interfere with subsequent analyses. BHT (butylhydroxytoluene, or 2,6-di-tert-butyl-p-cresol) is often employed because it elutes in or near the solvent front during GLC of long-chain fatty acid methyl esters.

(*ii*) Separation of Lipid Classes

The analysis of lipids frequently involves initial separation of the sample into the various component lipid classes that at the same time removes matrix interferences and concentrates the analyte. The separation can either be simple, yielding nonpolar and polar

fractions, or complex, resulting in separation and isolation of triacylglycerols, free fatty acids, sterols, steryl esters, glycolipids, acidic phospholipids, neutral phospholipids, etc. Traditionally, fractionation, cleanup, and concentration of lipids extracts have been achieved through the use of liquid extraction, thin-layer chromatography, or column chromatography, although other methods have also been employed. More recent methods include high-performance liquid chromatography and solid-phase extraction chromatography.

A. Column Chromatography (CC)

Column chromatography has been practiced for many years the traditional procedure to obtain purified lipid fractions in high amounts. It relies primarily on absorption and/or partitioning of the lipid components between the solid and the liquid (mobile) phases. Elution of the desired lipid classes is achieved by varying the polarity and strength of the mobile phase. Common stationary phases for CC of lipids includes: florisil, silicic acid, silica gel and alumina.

B. Thin Layer Chromatography (TLC)

Thin-layer chromatography is used extensively for lipid analysis and is a valuable tool for the separation and tentative identification of neutral and complex lipid classes. Although numerous sorbents are available for lipid TLC, silica gel is the one used most frequently. Prior to sample spotting, TLC plates can be treated with chemicals to alter its properties. In addition, TLC plates can be sprayed with or dipped in general or specific detection reagents for the identification of numerous lipophilic compounds. TLC also allows the removal of lipids from plates by scraping and extraction of compounds from the silica. Furthermore, the quantification of lipid classes is possible by in situ densitometry, and the use of two-dimensional (2-D) TLC is also va1uable for the separation of multi-component samples.

Numerous solvent systems are available for one-dimensional TLC of lipids. For neutral lipid separation, the system of Mangold (1984), or its modifications, is usually adequate.

C. Gas Liquid Chromatography (GLC)

Gas chromatography (or gas-liquid chromatography) still remains as the most valuable and efficient method of resolution and quantitation of fatty acids. This technique is based on the ideas of Martin and Synge, which was later applied by James and Martin to the separation of short chain fatty acids in the free form. Since then, the instrumentation has become more and more sophisticated with the development of new detectors, new columns, new injectors, and new ovens.

D. Solid-Phase Extraction (SPE) Chromatography

Solid-phase extraction is also very often used for lipid analysis because it is a simple and rapid technique that using the principles of traditional liquid-solid column chromatography and high-performance liquid chromatography, has a number of advantages over other sample separation methods. Thus, due to smaller sizes and volumes, SPE columns generally require less eluting solvents and disposal costs, resulting in direct cost savings. In addition, sample capacities and solvent elution volumes may be appropriate for direct injection onto a gas or liquid chromatograph without further sample preparation, reducing contamination and sample losses incurred during transfer steps. Furthermore, when sample concentration required, the small solvent volumes are evaporated easily and rapidly, and the traditional problems associated with emulsion formation common in liquid-liquid extractions, are eliminated with SPE

E. High-Performance Liquid Chromatography (HPLC)

The usefulness of HPLC for the separation of lipid classes is now well established, and the traditional problems associated with the detection of the nonchromogenic lipids in the column effluent have been largely overcome by the use of the universal light scattering, hydrogen flame ionization, and mass spectrometric detectors. Many different types of column packing materials, solvent elution systems, and detectors have been used for lipid class separations by HPLC, depending on the problem to be solved.

Normal-phase HPLC provides neutral lipid class separations of the type originally established for silicic acid absorption columns. Thus, triacyl glycerols, and mono- and diacyl glycerols are readily resolved from each other and from cholesterol esters and glycerophospholipids. A good separation of mono-, di-, and tri-acyl glycerols, 'free fatty acids, and cholesterol can be achieved in less than 15 min by using a column packed with silica gel and a gradient of ethanol into hexane-chloroform (9: 1, v/v) as elution system. Like TLC, normal-phase HPLC separations may also modified by inclusion of various modifiers in the adsorbent phase, such as silver ions to separate triacyl glycerols, based on the degree of unsaturation.

Although Gas Liquid Chromatography remains as the preferred technique for the analysis of fatty acids because of its very high resolution together with high sensitivity and very good reproducibility in quantitative analysis, it has also some disadvantages, particularly with respect to heat-labile fatty acids and the possibility of obtaining purified fatty acids in enough amount for further analysis. In order to overcome some of these shortcomings, a number of HPLC methods have been introduced.

These methods usually work at or near to ambient temperature and offer good resolution of the most important fatty acids.

Although fatty acids can be separated in the unesterified form by HPLC in the reversed-phase mode provided that an acidic compound is added to the phase, the detection of underivatized fatty acids is neither sensitive nor selective, because these compounds generally do not contain suitable chromophores. Therefore, many HPLC methods have been developed for derivatized fatty acids.

(*iii*) Fatty Acid Analysis Using Other Separative Techniques

Although gas chromatography and, to a lesser extent, high-performance liquid chromatography have been by far the most widely employed separative techniques for fatty acid analysis, other techniques have also been employed.

A. Capillary Electrophoresis (CE)

During recent years, CE has been developed into a powerful analytical technique for the separation of small molecules, such as pharmaceuticals, as well as for large biomolecules. It has also applied to fatty acid analysis with promising results. Nowadays, it is possible to separate most saturated and unsaturated fatty acids with a good resolution and employing very short separation times. On the other hand, the lack of suitable chromophores in the fatty acids imposes in many cases an indirect UV detection with a lower sensitivity.

B. Supercritical-Fluid Chromatography (SFC)

Supercritical-fluid chromatography is situated between GC and HPLC although it lacks the resolving power of either of them. It has been applied to fatty acid analysis. SFC has the advantage over GC of working at milder temperatures, facilitating the analysis of both thermally labile and nonvolatile compounds. In comparison to HPLC, SFC allows a wide range of detectors to be used, including the universal flame-ionization detector (FID), and an easier interfacing to mass spectrometry because of the smaller volumetric flow rates of the mobile phase going to the ion source. On the other hand, some technical deficiencies as well as an insufficient number of applications to real samples are observed at present.

(iv) Fatty Acid Characterization

(a) Chemical Procedures

1. Chain-Length Determination

Chain length may be easily determined by mass spectrometry or ^{13}C nuclear magnetic resonance spectroscopy. When using GC-MS, unsaturated fatty acids are converted into saturated fatty acids by hydrogenation and the saturated fatty acids formed may be easily identified. Hydrogenation is usually carried out in a test tube in the presence of hydrogen and platinum oxide as catalyst.

2. Location of Double Bonds

When using chemical procedures, the position of the double bonds is usually determined after cleavage of the fatty acid by oxidation and identification of the products formed. The two oxidation procedures generally used are those with permanganate-periodate or ozone as oxidative agents. The former method, which is the most largely used, yields mono- and di-basic acids as the products while ozonolysis followed by cleavage of the ozonides generates acids and aldehydes.

A typical protocol, recommended by Sempore and Bezard is the following: fatty acids are oxidized in tertiary butanol by sodium metaperiodate and potassium permagnate, with potassium carbonate as buffer for 1 h at room temperature. The solution is then acidified by sulfuric acid and the oxidative reagents are destroyed with sodium bisulfite. The acids formed are extracted with hexane and converted into sodium soaps with sodium methoxide to avoid loss of the short-chain acids. The solvent is evaporated and the soaps are converted into butyl esters by acidic butanol in a sealed tube at 100°C for 3 h. The chilled reaction mixture is treated with 5 per cent potassium carbonate and with pentane and shaken. After decantation the upper layer of butanol-pentane is directly used for GC analysis of the butyl esters, without evaporation

3. Location of other Functional Groups

Although spectroscopic methods are usually much more useful for these purposes, chemical methods can be an aid to spectrometric identification. Thus, epoxy groups are directly cleaved with periodic acid and the position of the oxygenated ring is deduced from the identity of the products formed. The ring of cyclopropane fatty acids can be disrupted by permanganate periodate oxidation and the ß-diketo compound produced can be identified by mass spectrometry. The ring of cyclopropane fatty acid reacts with boron trifluoride-methanol reagent to produce methoxy derivative which are 1ikely to be characterized by mass spectrometry. In branched fatty acids, the site of branding can be identified by GC by means of a keto compound formed by acidic permagnate oxidation.

(*b*) Spectroscopic and Spectrometric Techniques

In the study of the structure of fatty acids, spectroscopic and spectrometric methods have occupied for long time a very important position. There are a wide range of spectroscopic techniques employed ultra violet (UV), infrared (IR), nuclear magnetic resonance (NMR), and electron spin resonance (ESR)] as well as mass spectrometry (MS). In addition several of these spectroscopies can be subdivided. Thus, IR spectroscopy can be subdivided into mid and near infrared (MIR and NIR, respectively) and Raman spectroscopy, and, as described above, MIR may be coupled with GC. In addition, NMR spectroscopy encompasses high resolution (further split into ^{1}H and ^{13}C as the nuclei important in fatty acid analysis) and time domain or low resolution, and recent attempts at HPLC-NMR coupling have also been described.

(*v*) Estimation of Fatty Acid in a Meat Sample by Gas Liquid Chromatography (GLC).

Reagents and Materials

1. 10 per cent alcoholic KOH
2. HCl
3. Hexane
4. Distilled water
5. GLC with Flame Ionization Detector (FID)
6. GLC Columns*
7. Separating funnel
8. Pipettes
9. Funnels
10. Filter paper
11. Measuring cylinder

12. Micropipette
13. Vacuum Rotary Evaporator

* GLC columns are packed with 20 per cent diethylene glycol adipate (DEGA) on 101 chromosorb (60-80 mesh)

Principle

The lipid extract is converted into methyl esters and run into the GLC along with the standards and compared on the basis of retention time at the standard conditions.

Procedure

- ☆ The lipid extract was analyzed for the estimation of fatty acid by preparing fatty acid methylesters. Thirty ml of 10 per cent alcoholic KOH was added to 1 ml of lipid extract and allowed to stay in the dark overnight.
- ☆ Next day, the samples were saponified under reflux for about one hour. The contents of the flask were cooled and transferred into separating funnel.
- ☆ Thirty ml of distilled water and 50 ml of hexane were added. After thorough mixing, the two phases were allowed to separate.
- ☆ The lower phase was collected in the same flask and the upper phase, which contain non-saponifiables were discarded.
- ☆ To the lower phase in the flasks 15 ml HCl and 50 ml of Hexane were added, mixed and allowed to separate in a separating funnel. Hexane (the upper layer) was collected in a separating flask.
- ☆ The lower phase was collected in old flask and again washing was given with 30 ml distilled water and 50 ml hexane, mixed and allowed to separate.
- ☆ Hexane layer was collected in the hexane flask and lower phase was discarded. Hexane collected was evaporated to dryness in the rotary evaporator.
- ☆ To the vacuum dried fatty acid samples, 50 ml of 3 per cent methanolic HCl was added and refluxed for 1 hr.
- ☆ Few glass chips were added to avoid bumping. Flasks were removed and contents were transferred to a

separating funnel to which 30 ml of distilled water and 50 ml of hexane were added and mixed and allowed to separate.

- ☆ The lower phase was collected into the same flask and hexane layer was collected in a separate flask.
- ☆ The procedure was repeated twice and the collected hexane was evaporated and dried upto 2 to 3 ml.
- ☆ Final concentration of methyl esters in hexane was 1 mg/ml. The samples of methyl esters were stored at –18°C.
- ☆ Fatty acid methyl esters were analyzed isothermally by Gas Liquid Chromatography equipped with Flame Ionization Detector (FID) and columns packed with 20 per cent diethylene glycol adipate (DEGA) on 101 chromosorb (60-80 mesh). The operating conditions should be followed:

  Column temperature = 173°C

  Detector temperature = 220°C

  Injection temperature = 240°C

  Sample size = 1µl

The fatty acid peaks were identified by comparing the retention time of standard fatty acids.

(*vi*) Extraction of Lipids from Raw and Cooked Meat Samples

The method of Folch *et al.* (1957) is still most acceptable method for the extraction of lipids from raw and cooked meat samples.

Reagents and Materials

1. Chloroform
2. Methanol
3. Sodium chloride solution
4. Butylated Hydroxy Toulene (BHT)
5. Separating funnels
6. Rotary vacuo Evaporator
7. Whatman No. 1 Filter papers

8. Funnel
9. Weighing balance

Procedure

- ☆ Take 5 g of sample and ground in a pestle and mortar with 20 volumes of solvent mixture comprising of chloroform: methanol (2:1, v/v).
- ☆ Allow the content to stand at room temperature with occasional stirring for 6-8 hours.
- ☆ Filter the extract through Whatman No. 1 filter paper and re-extract the residue with 10 more volumes of the same solvent mixture for 2 hours and filter it.
- ☆ Dry the filtrate in vacuum at 55-60°C in a rotary evaporator.
- ☆ For breaking the proteo lipids, dried lipid residue was dissolved in one tenth volume of the original lipid extract in chloroform: methanol: water (64: 32:4, v/v/v) and evaporated to dryness in vacuum at 55-60°C.
- ☆ Repeat this step twice and dissolve the dried lipid residue and filter quantitatively into a separtory funnel with 100 ml of chloroform : methanol (2:1, v/v).
- ☆ Wash the lipid extract with one-fifth volume of 0.9 per cent sodium chloride solution so as to remove non-lipid impurities from the lipid sample.
- ☆ It was allowed to stand overnight at room temperature.
- ☆ The chloroform layer was collected and evaporated to dryness in vacuum at 55-60°C and volume was made upto 10 ml with chloroform.
- ☆ The lipid samples were stored in a stoppered glass test tubes and to each tube a drop of 0.5 per cent Butylated Hydroxy Toulene (BHT) in chloroform was added as preservative.
- ☆ These samples were stored in a deep freeze at –18 °C till further analysis.

(*vii*) Estimation of Total Lipids by Gravimetric Method

Reagents and Materials

1. Aliquot of lipid extract

2. Hot air oven
3. Weighing balance
4. Stainless steel planchet

Procedure

The total lipids of the samples were determined gravimetrically by a method described by Bligh and Dyer (1959).

- ☆ One ml of aliquot of the lipid extract was pipetted into a dried stainless steel planchets with constant predetrmined weights.
- ☆ The samples were dried at 60°C in a hot air oven to a constant weight.
- ☆ Total lipids were expressed as mg/g of a sample.

(*viii*) Estimation of Total Phospholipids

Principle

Total phospholipids were estimated by determining the phosphorus content of the lipid extract by converting into inorganic phosphorus. The modified method of Marinetti (1962) is used for the estimation of phospholipid phosphorus.

Reagents and Materials

1. Lipid extract
2. Potassium dihydrogen phosphate
3. Distilled water
4. Ammonium molybdate
5. ANSA (1 Amino 2- Naphthol 4 Sulfonic acid)
6. Micropipette
7. Spectrophotometer
8. Hot water bath
9. Digestion bench
10. Measuring cylinder

Procedure

- ☆ Twenty five micro litre lipid extract and standard phosphate solution (25, 50, 75 and 100 μl) containing 0.4

mg/5 ml potassium dihydrogen phosphate were digested with 1 ml of 60 per cent perchloric acid.

- ☆ Blank was prepared with perchloric acid only and treated in a similar fashion.
- ☆ After digestion, 0.5 ml of 2.5 per cent ammonium molybdate solution was added followed by 0.2 ml of ANSA reagent (1 Amino 2- Naphthol 4 Sulfonic acid).
- ☆ After thorough mixing 7.0 ml distilled water was added and tubes were heated in water bath at 55-60°C for 7.0 min.
- ☆ The color developed due to formation of molybdenum blue was measured by reading the optical density at 830 nm in Beckman DU-40 Spectrophotometer.
- ☆ The inorganic phosphorus content was multiplied by a factor of 25 to arrive at phospholipid content and expressed as mg/g of sample.

(*ix*) Estimation of Total Cholesterol

Principle

Total cholesterol in the lipid extracts were determined by adopting the method described by Hanel and dam (1955).

Reagents and Materials

1. Lipid extract
2. Chloroform
3. Distilled water
4. Acetic acid
5. Sulphuric acid (H_2SO_4)
6. Micropipette
7. Spectrophotometer
8. Hot water bath
9. Digestion bench
10. Measuring cylinder

Procedure

- ☆ Twenty five µl of lipid extract of meat sample and 50 µl standard solution of cholesterol of concentration

1 mg/ml was pippetted out in test tubes and kept in water bath 55-60°C to evaporate chloroform.

- ☆ Add 3 ml of acetic acid in dried test tubes.
- ☆ Blank was prepared with acetic acid only and treated in a similar fashion.
- ☆ Then add 0.1 ml distilled water and 2 ml $FeCl_3$ in acetic acid. To prepare working solution, take 1 ml of stock solution and dilute it to 100 ml with concentrated H_2SO_4.
- ☆ All the reagents were mixed well and kept for half hour for color development.
- ☆ The optical density was measured at 560 nm in a spectrophotometer and expressed as mg/g of tissues.

(*x*) Estimation of Glycolipids

Principle

Total glycolipids were estimated by determining the hexose (galactose) content of lipid extract. A method of Roughan and Batt (1968) is commonly used for estimation of galactose in lipid extracts.

Reagents and Materials

1. Lipid extract
2. Chloroform
3. Phenol
4. Galactose standard solution
5. Conc. sulphuric acid
6. Micropipette
7. Spectrophotometer
8. Hot water bath
9. Digestion bench
10. Measuring cylinder

Procedure

- ☆ The galactose standard solutions of 20, 40, 60, 80 µl of concentration 1mg/1ml of distilled water was taken in test tubes.

- ☆ One ml of 2 per cent phenol along with 4.0 ml of concentrated H_2SO_4 was added.
- ☆ Orange color appeared was measured at 480 nm after cooling to room temperature for 15 min.
- ☆ The standard curve was plotted, taking concentration on X-axis and O.D. on Y-axis.
- ☆ One ml of lipid sample was evaporated to 2 hours at room temperature (20-25 °C).
- ☆ After hydrolysis, 4.0 ml of chloroform was added and the mixture was centrifuged.
- ☆ Take out 1 ml of aqueous layer and 1 ml of 2 per cent phenol followed by 4 ml of concentrated H_2SO_4.
- ☆ The orange color developed was measured at 480 nm in a spectrophotometer.
- ☆ The concentration of galactose of the lipid was calculated from the standard curve.
- ☆ This was multiplied by 4.45 to estimate the glycolipid content and expressed as mg/g of sample.

(*xi*) Estimation of Free Fatty Acid (FFA)

Principle: The method used for the estimation of FFA is Koniecko, 1979.

Reagents and Materials

1. Lipid extract
2. Methanol
3. 0.1N NaOH
4. Phenolphthalein indicator
5. Palmitic acid
6. Pipette
7. Burette

Procedure

- ☆ One ml of lipid extract was taken in 5.0 ml of freshly neutralized methanol.
- ☆ Fatty acids were titrated against 0.1 N NaOH using phenolphthalein indicator.

- ☆ Palmitic acid was used as standard FFA, content was expressed as mg/g of sample.

(*xii*) Estimation of Total Glycerides

Procedure

- ☆ Total glycerides are indirectly calculated by subtracting the sum of total phospholipids, total cholesterol, total glycolipids and total free fatty acids from total lipid values.

Chapter 7

Microbiological Quality

1. Orientation and Use of a Compound Microscope

A compound microscope is defined as a magnifying optical instrument consisting of two lens systems, the objective and the ocular.

Parts of a Microscope

Base, arm, inclination joint, mechanical clips, condenser with diaphragm, mirror or light source, body tube, revolving nose piece, objectives, draw tube, coarse and fine adjustments and eye piece.

Magnifying Power of a Microscope

Magnifying power of the objective × Magnifying power of the eye piece.

A. Use of Low Power Objective

(*i*) Preparation of a Wet Mount

Take a clean slide (Clean the slide by washing with detergent powder and wiping with a clean piece of cloth. Make it grease free by passing over the flame for two to three times). Place a drop of the suspension on the glass slide. Place the cover slip over the drop taking care to avoid any air bubbles. Set the microscope on the table with the mirror facing the lamp or outside light.

(*ii*) Preliminary Focusing and Use of Low Power Objective

1. Arrange the microscope and the mirror so that the light coming up from the mirror passes through the slide.
2. Place the wet mount on the stage of the microscope.
3. Swing the low power objective into position. Make certain that it does not touch the cover slip.
4. Turn the coarse adjustment until the object appears in approximate focus. The distance between the bottom lens of the objective and the object when in focus is called the working distance. Higher the magnification, smaller the working distance.
5. Use the fine adjustment to sharpen the focus. Once the object is seen, do not use the coarse adjustment.
6. Draw a picture of the specimen as it appears under the low power objective and mention the magnification.
7. Use of condenser and mirror: Proper lighting is of primary importance to successful microscopy and this can be obtained by appropriate adjustment of the condenser and the mirror. Move the condenser all the way down slowly. Describe the effect of moving the condenser.
8. While looking through the eye piece move the metal handle which adjusts the diaphragm and get familiar with the functions of Iris diaphragm.

B. Use of Medium Power Objective (Microscopic Examination of Protozoa, Algae and Yeast)

Protozoa

These are unicellular microorganisms which are non-photosynthetic, and mostly motile.

Algae

These are unicellular or multicellular, photosynthetic, and motile or non-motile microorganisms.

Yeast

They are unicellular fungi, divide mostly by budding.

Procedure

- ☆ Prepare a wet mount of the given samples on a clean slide.
- ☆ First examine under low power objective and describe the object.
- ☆ Swing the medium power in position taking care that it does not touch the cover slip and then sharpen the focus with fix adjustment knob only.
- ☆ Make the labelled diagram. Mention the magnification of the object.

C. Use of Oil Immersion Objective

Microscopic examination of bacteria and Actinomycetes.

Principle

In low and medium power objectives, the light passes from denser medium (glass) to a rarer medium (air) and so refracts away from its path. By placing a drop of oil whose refractive index is equal to that of glass, more light is made to pass through the object to the objective with the result the object will be seen brighter. The resolution will also be more.

Resolving power of the microscope is the ability of the objective to differentiate between two objects, which are placed very near to each other, and it depends on the wavelength of the light used for observing the object.

Preparation of Slide of Bacteria (Simple Staining)

Procedure

- ☆ Take a clean and grease free slide and prepare a smear of the given culture suspension by placing one drop on the slide and spread it uniformly, with the help of inoculating needle and allow it to air dry.
- ☆ Fix the smear by passing the slide over the flame 2 to 3 times. Bacteria will fix to the slide.
- ☆ Stain with crystal violet solution for one minute and wash gently with tap water and dry.

For Use of Oil Immersion Objective

Procedure

- Mount the slide on the microscope.
- Using low and medium power objectives get the object in focus. Describe how you see the object under each objective.
- Place a drop of cedar wood oil over the centre of the slide and raise the condenser close to the slide.
- Carefully turn the oil immersion objective (the objective should touch the oil) and look through the ocular.
- Turn the fine adjustment slowly till the object is in focus.
- Close the iris diaphragm slightly so that light continues to fill the lens.
- This lighting will give maximum contrast between the object and background.
- Describe the object as seen under the oil immersion objective and make drawings.

Precautions

- Before returning the microscope to the cabinet move the low power objective into place at least ½″ above the stage.
- Wipe the oil off the lens with a clean lens paper with xylene and wipe the lens.
- Remove the residual xylene with a dry place of lens paper.

2. Use of Common Bacteriological Apparatus

(*a*) Hot Air Oven

Double jacketed, dry air circulated with thermostatic control, glasswares are sterilized using hot air oven at 160°C for 2.5h or at 180°C for 1.5 h. Temperature of 160°C is preferred than 180°C to prevent charring.

(*b*) Autoclave

Used for sterilization of media, rubber and plastic wares. The moist heat at 15 lbs pressure or 121°C for 15 min. is required for sterilization. Moist heat has better heat penetration power to kill spore formers.

(*c*) Incubator

Range of temperature is 25 to 100°C. Used for incubating the Petri plates inoculated with test samples, operated at 35°C for 24-48h for APC or SPC. It is thermostatically controlled.

(*d*) BOD Incubator

It is with temperature and humidity control. It can be used in wider range of temperature, from refrigerating or chilling temperature to higher temperature upto 100°C

(*e*) Colony Counter

Automatic or Quebec colony counters are used.

(*f*) Microscope

Used to study the microbiological characters, staining characters of bacteria. Two types Compound and Electronic microscopes.

3. Cleaning and Sterilization

Cleaning and sterilization of glasswares are of utmost importance to obtain accurate microbial analysis

- ☆ Cleaning is done by using warm (50-55°C) detergent solution (1 per cent washing soda or trisodium phosphate) using a brush.
- ☆ For cleaning the sticky materials from pipettes, these are immersed in chromic acid solution (80g Potassium dichromate in 300 ml water + 460 ml of concentrated sulphuric acid) for 6-8h before detergent cleaning.

For sterilization, following procedure may be followed:

- ☆ Petri plates and pipettes are kept in respective containers before putting in a hot air oven
- ☆ Flasks and bottles can be directly kept inside the oven.
- ☆ In case of glass stoppered bottles, a piece of paper should be inserted in the mouth of the bottle prior to sterilization to prevent jamming.
- ☆ After putting the oven on, wait for it to achieve a temperature of 160-180°C. Maintain this temperature for 2-2.5h. After this, gradually cool down for about 2h.

☆ Do not open the oven immediately after switching off, as it causes serious breakage.

4. Gram's Staining Technique

Bacteria stained by Gram's method, fall into Gram +ve and Gram–ve group. Gram +ve bacteria retain the crystal violet color and hence appear deep violet or blue in color. Gram-negative bacteria loose crystal violet color when treated with a decolorizer and are counterstained by Safranin and hence appear red in color. The gram staining may be explained in two ways.

The cell wall of Gram–ve bacteria is thinner than Gram +ve bacteria and contains high proportion of lipid. Alcohol treatment extracts the lipids and increases the permeability of cell wall, which extracts the crystal violet iodine complex and hence takes the color of counter stain.

Gram-ve bacteria have a very much smaller amount of peptidoglycan, which is less extensively cross-linked than that in the walls of Gram +ve bacteria. The pores in the peptidoglycan of Gram–ve bacteria remain sufficiently large even after ethanol treatment to allow the crystal violet iodine (CV-I) complex to be extracted.

There are many bacteria, which are Gram variable rather than true Gram +ve or Gram–Ve. The old culture of some Gram +ve is easily decolorized and therefore takes the red color of counter stain.

Composition of Gram's Stain

(A) Ammonium Oxalate Crystal Violet (Hucher's)

Solution I	*Solution 2*
Crystal violet: 2 g	Ammonium oxalate: 0.8 g
Ethyl alcohol 95 per cent: 20 ml	Distilled water: 80ml

Mix solution 1 and 2 and then filter it.

Lugol's Iodine Solution

Iodine: 1g

Potassium iodide: 2 g

Distilled water: 300 ml

Dissolve the ingredients and filter it.

Ethyl alcohol 95 per cent (decolorizer)

Safranin (counter stain)

Safranin 'O' (2.5 per cent solution) in 95% ethyl alcohol: 10 ml

Distilled water: 100 ml

Dissolve the stain and filter it.

Procedure

- ☆ Prepare a thin smear of bacterial culture in a grease free slide and allow it to dry. Heat the smear to fix it.
- ☆ Stain the smear with crystal violet for 2 min and wash it in the tap water.
- ☆ Add Lugol's iodine solution and allow it to act for 1 minute.
- ☆ Decolorizes in 95 per cent ethyl alcohol for about 30 seconds gently agitating the slide till no more color comes out from the smear.
- ☆ Counter stain with Safranin for 10-20 seconds.
- ☆ Wash in tap water, dry and examine in microscope under oil immersion.

Interpretation

Gram +ve bacteria will be deep violet color and Gram–ve bacteria will be red in color.

5. Spore Staining and Demonstration of Capsule of Bacteria

Bacteria belonging to genera *Bacillus, Clostridium* and *Sporosarcina* are characterize by their ability to form heat resistant resting structures called endospores.

Reagents and Materials

1. Malachite green
2. Safranine
3. Glass slide
4. Microscope

Procedure

Malachite Green Staining

- ☆ Prepare a smear from the sporulated culture as before.
- ☆ Add two drops of 6 percent Malachite green and warm the slide over a flame for two minutes. Do not allow the slide to dry.
- ☆ Rinse with water for 10 seconds.
- ☆ Counter stain with 0.25 per cent aqueous safranine for 30 seconds.
- ☆ Rinse, blot dry and examine under oil immersion objective.
- ☆ Make diagram of the bacterial spore.

Results

Pink colored cells containing green spores inside the cell. Some will be seen as free spores.

Capsule

Slide of capsule staining will be shown in the class. Violet colored cells surround by faint blue to colorless capsules.

6. Microbiological Sampling of Meat, Poultry Products and Eggs

Meat and meat products are prone to be contaminated with one or another type of bacteria during various processing steps. Generally, total bacterial load is estimated in meat products to know the sanitary condition under which meat is produced, processed and marketed.

Properties of Good Sample

1. Ideal sample should be identical in all properties with the bulk of material.
2. It should be large enough for all types of subsequent, intended determinations.
3. Size of samples varies from material and type of analysis.
4. It should be packed and stored in such a way, that no significant change occurs from moment of sampling till analysis.

Microbiological Sampling of Meat/Poultry Products

A. Fresh dressed meat/poultry: Sample may be drawn by any of the following methods:

- *Swab method*: A sterile cotton swab is immersed in sterile saline or peptone water and applied over the surface of meat on specific portions *e.g.* $1cm^2$, $2cm^2$, $4cm^2$ etc at different places. After swabbing the surface, dip the swabs back in sterile diluent solution and break the stick at about ½" above the swab. Shake the tubes vigorously to discharge the organisms.
- *Triturate method*: Samples are taken from different part of meat by using sterile knife or scissors. Generally, 10 g of meat is cut using sterile knife and meat sample is blended with sterile saline using sterile pestle and mortar.
- *Direct rinse method*: Place the meat portion/dressed poultry in a sterile polyethylene bag containing sterile diluents (1:10). With the help of a stomacher the meat sample is homogenized to make a meat homogenate. Known volume of this homogenate is used for microbial evaluation.

A. Meat/Poultry products

The microbiological procedures can be followed as mentioned earlier but some specific precautions should be taken for different type of meat and meat products.

- *Frozen meat*: Frozen sample is rejected if it reaches the laboratory in thaw condition because microbial load might have increased during transition. Thaw the frozen sample in a water bath at 45°C for not more than 30 minutes or thaw at room temperature for 2-4 hours. Never refreeze the sample. Drip collected during thawing can also be used for microbial analysis.
- *Cooked meat*: Remove the sample in cooked meat as in fresh meat. Generally, large area for swab is used due to lower microbial load in cooked meat.
- *Deboned meat*: Any of the earlier method can be used.
- *Canned meat*: Examine the cans for abnormalities such as flipper, soft swell, hot swell and leakage etc. Remove these abnormal cans and determine microbial loads and isolate

the microbes. Incubate normal cans at 35°C for 10 days and examine daily. In case cans do not develop any change till 10 days terminate the analysis. Before opening cans heat the area with a burner and open with pasteurized device. Transfer contents immediately to selected media under laminar flow chamber. Do not flame swells as they may burst. Instead use chlorine solution.

B. Egg and Egg Products

- *Shell Eggs:* Shell eggs are collected randomly from a lot and these are packaged cleanly either in carton or filler flat trays with proper identification markings. The temperature of the eggs should be maintained below 50 °F till the analysis. Preferably, the eggs should be kept at room temperature 18 ± 2°C to avoid the sweating of eggs. As sweating increases the chances of contamination of egg content.
- *Whole Egg Liquid*: Thoroughly mix the content of container either mechanically or manually. Avoid frothing. Fill the sample in a clean, dry, fat free, sterile container with the help of dipper. Fill the can ¾ of its capacity and keep the temperature below 5°C but avoid freezing. Analyze the sample with in 4 hours of sampling. Observe and record odour of each container sampled before analysis. Mix the content thoroughly and collect the sample with sterile spoon. Prepare serial dilution and perform microbiological assays.
- *Frozen Whole Egg Liquid*: Randomly select a can from a lot. Remove a top layer of egg liquid with sterilized hatchet or chisel. Drill three cores from top to bottom of container: first core in centre, second core midway between centre and periphery and third core near edge of container. Collect all the samples from cores into a sample container with the help of sterile core plunger or spoon. Refrigerate samples with solid CO_2 or other suitable refrigerant till analysis. The sample collected aseptically as described earlier is thawed by submerging the holding sample containers into cold water or with the help of microwave. Mix the sample content thoroughly with a sterile glass

rod and take the measured amount for the preparation of serial dilution and subsequent plating for microbiological analysis.

- ☆ *Dried Eggs*: For small packages, take entire parcel for sample. For boxes and barrels remove top layer with sterile spoon and with sterile trier, remove three cores as mentioned above. Aseptically transfer core to sample container with sterile spoon or other suitable instrument. Store samples under refrigeration till analysis.

Precautions

(*i*) Label all the samples.

(*ii*) The temperature difference between soapy water and egg should not be more than 20 °F.

(*iii*) Be precautious while flaming off excess alcohol.

(*iv*) Take out sample completely from the corer.

(*v*) Avoid foaming while mixing egg material with diluent. If foaming occurs than stop mixing. Keep it for few minutes to settle down and then again start mixing.

7. Preparation of Homogenate and Decimal Dilution

Objective

- ☆ To prepare homogenous mixture of meat sample supplied and to prepare ten fold dilutions of this homogenate for microbiological analysis.

Significance

There are several products prepared out of meat for human consumption. The examination of these foods for the presence, types and number of microorganisms and/or their products is the basis of food microbiology. These microorganisms grow in these products and cause spoilage and transmit diseases. Hence, examination of foods for microbes is of great public health significance.

As microbiological analysis of total food products cannot be possible as such, the representative samples are taken and made into suitable homogenous solution.

Principle

When the homogenous mixture of food or food/meat is prepared. The microorganisms present in the food distributes throughout the mixture equally along with the solute materials. Thus we get uniform microbial mixture. If we dilute the mixture by 10-fold method, we can get the uniform dilution also

Procedure

- ✰ Ten gram of meat sample is taken, macerated to which 90 ml of 0.1 per cent peptone water is added to make the volume 100 ml. The mixture is properly blended to make a dilution as 1:10 (10^{-1}).
- ✰ Of this 1 ml is taken from the above mixture and transferred into test tube containing 9 ml of peptone water. Now the dilution of the test tube is 1:100 (10^{-2})
- ✰ 1ml of 10^{-2} dilution mixture is again transferred into another test tube containing 9 ml peptone water to get the resultant dilution as 1: 1000 (10^{-3})
- ✰ This procedure is repeated further to get dilutions of 10^{-4}, 10^{-5}, and 10^{-6} and so on.
- ✰ These dilutions of the mixture are made in 10-fold manner, so it is called decimal dilution. Thus the microbial population decreases uniformly and proportionately as the dilution increases.

8. Isolation and Cultivation of Aerobic Microorganisms

In nature, microorganisms are found in mixed population, and therefore if one is to study a single type of organism it is necessary to have a pure culture. A pure culture is defined as that which contains single type and that has developed from a single cell. Once the pure culture is obtained it can be grown and its morphology and staining characteristics are determined.

Several methods are available for separating microorganisms from one another. These include both agar and broth dilution procedures and the use of selective differential media. In order to obtain satisfactory results precautions must be taken during the performance of all procedures to avoid air currents created by open windows, heating system etc. To achieve this, isolation should be done in inoculation chamber.

The methods used in the isolation of pure cultures of aerobic bacteria are:

- ☆ Pour plate technique.
- ☆ Streak plate technique.

(*a*) Pour Plate Technique

Procedure

- ☆ Make serial dilution of the given sample.
- ☆ Take two sterile Petri dishes and write your name and date.
- ☆ Add 1 ml of the last serial dilution to the sterile plate.
- ☆ Cool the agar medium to 45°C. Remove the plug near to the flame and pour about 15-20 ml of the medium in the Petri plate containing 1 ml of sample.
- ☆ Rotate the Petri plate gently to mix the sample with medium.
- ☆ After the agar has solidified, incubate the Petri dishes upside down at the appropriate temperature.

Make the observations next day and record.

(*b*) Streak Plate Technique

Procedure

- ☆ Melt and cool the agar medium to 45°C as before and pour into sterile petri plates. Allow to cool and solidify.
- ☆ Remove a loopful of the culture suspension aseptically.
- ☆ Lift the top of the Petri plate and place the inoculums at the edge of the agar farthest from you and streak (as shown in diagram no. 1).
- ☆ Incubate the plates at appropriate temperature. The first streak will contain most of the culture while the last streak will thin out the culture sufficiently to give isolated colonies.
- ☆ A pure culture may be maintained by transferring a portion with an inoculating needle as described below:

(*c*) Transfer and Maintenance of Micro-organisms

Procedure

- ☆ Sterilize the inoculating needle and allow it to cool.
- ☆ Transfer a portion of an isolated colony from each of the two Petri dishes to agar slants aseptically.
- ☆ Incubate the slants at appropriate temperature for 24 to 48 hours.
- ☆ Prepare a smear as discussed earlier from the culture growing on the slants.
- ☆ Stain by Gram's procedure to ascertain the purity of the culture.

(*d*) Maintenance

After growth on slant, the cultures are to be maintained by lowering their metabolic activities. The different methods are:

- ☆ By storing the cultures in refrigerator at 4-8°C. Don't put the culture into the freezer because freezing kills micro-organisms. Subculture from old to new slants after every two months.
- ☆ Storing under sterilized mineral oil. In order to avoid frequent sub-culturing, sterilized mineral oil is added to the slant to cover the whole slope of the medium. Mineral oil prevents the drying of the medium and culture can be maintained viable for 6-12 months in a refrigerator.

9. Demonstration of Ubiquitous Nature and Growth of Micro-organisms

Microorganisms are ubiquitous in nature:

- ☆ Pour 10-15 ml of nutrient agar medium (cooled to 45°C) aseptically in a sterile petriplate.
- ☆ Allow the medium to solidify.
- ☆ Expose these petriplates to different ecosystems (expose it to air for a second, put your finger, corner of handkerchief, a few drops of drinking water, biogas fermenting broth, sewage water, saliva, teeth scrapings, soil, curd, fruit and vegetable washings, etc.).

- ☆ Put one plate unopened as control.
- ☆ Incubate the Petri plate at 30°C for 48 hours.
- ☆ Record your observations and confirm that unopened plate does not develop any colony where as all other ecosystems are laden with micro-organisms.

Growth of Micro-organisms

Growth in unicellular microorganisms (bacteria) is manifested by increase in the cell counts and this result in the increased turbidity of the medium. In multicellular organisms (fungi) growth results in increase weight of the mycelium.

Procedure

Bacterial Growth

- ☆ Inoculate aseptically a flask containing broth with a loopful of freshly grown bacterial culture.
- ☆ Determine the turbidity with the help of colorimeter at 540 nm after 6, 12, 18 and 24 hours.
- ☆ Determine the relationship of turbidity with time.

Fungal Growth

- ☆ Inoculate aseptically a series of flasks containing Czapak's broth with the spores of *A. nigar*.
- ☆ Take out fungal mats after 12, 24, 48 hours of growth.
- ☆ Dry the mycelial mat 80°C for 24 hours.
- ☆ Weigh the mycelial mats and find out the relationship between weight and time.

10. Total Plate Count

The total plate count has very limited usefulness in meat quality control since it is not related to flavor until it reaches about 10^7 organisms per gram, and tells us little about keeping quality. The total plate count is useful, however, for assaying the total sanitary effect of processes, for preparing flow sheets of bacterial contamination, and in similar cases. Often, in such cases, most probable number (MPN) technique is just as useful as plate counts, and a decision on which method to use is largely a question of the operator's preference. If one is interested only in organisms of public health significance, and concomitantly, in determining counts in connection with the work of regulatory agencies, 35°C is a good incubation temperature. Where one is concerned with keeping quality of products, which are refrigerated, either 20 or 25°C should be used. Plates held at 35°C are counted after 24 h and sometimes held for a second count at 48 h. Plates incubated at 20 or 25°C can hardly be counted before 3 days, and 5 days is often better.

Standard Plate Count (SPC)

Media: Tryptone Glucose Yeast Extract Agar or Plate Count Agar

Ingredient	*Quantity*
Tryptone	5.0 g
Yeast extract	2.5 g
Glucose	1.0 g
Agar	15.0 g
Distilled Water	1.0 litre
pH	7.1 ± 0.1
Autoclave at 121°C for 15 min	

Procedure

- ✩ From a thoroughly mixed ground meat sample, aseptically prepare, a 1:10 dilution by weighing 10 g of meat into sterile dilution bottles containing 90 ml of either sterile 0.1 per cent peptone water or 0.85 per cent physiological saline (NaCl)
- ✩ Thoroughly agitate the bottle by shaking rapidly 25 times in a 1 foot arc within 7 sec. Dried samples may require more agitation.
- ✩ Prepare subsequent serial dilutions as required.
- ✩ Transfer 1 ml from appropriate dilution to the Petri dishes (15×100mm) in duplicate.
- ✩ Add approximately 15 ml prepared plate count agar (Tryptone-Glucose yeast extract agar) to the Petri dishes at about 45°C.
- ✩ Thoroughly mix and allow to solidify.

- ☆ Incubate the inverted plates at 35°C for 48±2h.
- ☆ Count the colonies and report the result as standard plate count as cfu/g after multiplying the average number of colonies with dilution factor.

11. Determination of Coliform Count

Presence of coliform organisms, particularly the bacterium *Escherichia coli* is generally regarded as an indication of faecal contamination. In case of meat products, which are heated to pasteurizing or higher temperature during processing, the presence of Coliforms may be a good measure of post processing contamination.

Most Probable Number (MPN) Method

(*i*) Presumptive Coliform Test

Media: Lauryl Sulphate Tryptose (LST) broth

Ingredient	*Quantity*
Tryptose or Trypticase	20.0 g
Lactose	5.0 g
Dipotasium Phosphate	2.75 g
Monopotassium Phosphate	2.75 g
Sodium Chloride	5.0 g
Sodium Lauryl Sulphate	0.1 g
Distilled Water	1.0 litre
pH	6.8 ± 0.2
Autoclave at 121°C for 15 min	

Procedure

- ☆ Prepare sterile test tubes containing 10 ml of lauryl sulphate tryptose (LST) broth with an inverted fermentation tube.
- ☆ Incubate each of 3 LST tubes with 1 ml of undiluted sample of meat. Inoculate 3 additional LST tubes using 1ml of a 1:10 dilution of the sample and three tubes using 1ml of 1: 100 dilutions.

- ☆ Incubate at 35°C for 48 ± 2h.
- ☆ Examine (for gas production) at 24h and if negative again examine at 48 h.

Perform a "Confirmed test" on all presumptive positive (gassing) tubes.

(*ii*) Confirmed Tests for Coliforms

Media: Brilliant Green Lactose Bile Broth, 2 per cent

Ingredient	Quantity
Peptone	10.0 g
Lactose	10.0 g
Ox gall or Bile extract	20.0 g
Brilliant Green	0.0133 g
Distilled water	1.0 litre
pH	7.2 ± 0.2
Autoclave at 121°C for 15 min	

Procedure

- ☆ Gently agitate each gassing LST tube and transfer a loopful of the suspension to a tube of brilliant green lactose bile (BGLB) broth (2 per cent).
- ☆ Incubate 48±2hrs at 35°C.
- ☆ Examine for gas production at 24 and 48 hrs and record results.
- ☆ Using "Most Probable Number (MPN)" table, complete the MPN of Coliforms on the basis of tubes of BGLB broth, which exhibit gas production.
- ☆ Use MPN test when coliform count is expected to be less than 110/g.

(iii) Plate Count Method for Coliforms

Media : Violet Red Bile (VRB) Agar

Ingredient	Quantity
Yeast Extract	3.0 g
Peptone	7.0 g

Ingredient	Quantity
Sodium chloride	5.0 g
Lactose	10.0 g
Bile salts	1.5 g
Neutral Red	0.03 g
Crystal violet	0.002 g
Agar	15.0 g
Distilled water	1.0 litre
pH	7.4 ± 0.2
Don't sterilize, boil for 2 minutes	

Procedure

- ✰ Prepare dilutions beyond the 1:100 dilutions as needed. Counts should be made on plates with 15 to 150 colonies
- ✰ Prepare duplicate plates for each dilution using 1ml of the diluted meat sample.
- ✰ Add 15 g of Violet Red Bile (VRB) Agar into each Petri plate.
- ✰ When the agar was solidified, pour a cover over the cooled agar using about 3-4 g of VRB agar.
- ✰ When cool, invert and incubate for 24±2 hr at 35°C.
- ✰ Count as Coliforms only those dark red colonies, which are at least 0.5 mm in diameter on uncrowded plates.
- ✰ Report as Coliforms per gram of product.

(*iv*) Presumptive *E. coli* Determination

Media: EC Broth

Ingredient	Quantity
Pancreatic digestion of casein	20.0 g
Bile salts	1.5 g
Lactose	5.0 g
Dipotassium phosphate	4.0 g
Sodium chloride	5.0 g
Distilled water	1.0 litre
pH	6.9 ± 0.2
Autoclave 15 min. at 121 °C	

Media: Eosin Methylene Blue Agar (Levine)

Ingredient	Quantity
Ingredient	Quantity
Peptone	10.0g
Lactose	10.0g
Dipotassium phosphate	2.0 g
Methylene Blue	0.065 g
Agar	15.0 g
Distilled water	1.0 litre
pH	7.1 ± 0.1
Autoclave 15 minutes at 121°C	

Procedure

- ✰ Follow the procedure outlined for Coliforms.
- ✰ For each LST tube showing gas production prepare a sterile test tube containing 8 ml of EC broth and on inverted fermentation tube. Transfer a 3mm loopful of LST broth to the EC broth tube
- ✰ Incubate in a covered water bath 48±2 hrs at 45.5 ±0.05 °C.
- ✰ Examine for gas production at 24±2 hrs and if negative, again at 48 ± 2hrs.
- ✰ Streak loopful of suspension from each gassing tube to Levine's Eosin Methylene Blue (L-EMB) agar.
- ✰ Incubate 18 to 24 hrs at 35°C.
- ✰ Examine for suspicious *E. coli* colonies that are dark centered with or without metallic sheen.
- ✰ Report as presumptive *E. coli.*

12. Determination of Yeast and Mould from Meat

Yeast and moulds are responsible for spoilage of many types of foods. Yeast and moulds often manifest themselves on or in foods in which conditions are less favorable to bacterial growth. These are foods of low pH, low moisture, high salt or sugar content, or under conditions such as low storage temperature, the presence of antibiotics, or following exposure of food to irradiation. Yeast and moulds can utilize such substances as pectin and other carbohydrates, organic acids, proteins and lipids. Yeast and/or moulds can cause problems through

- Synthesis of toxic metabolites
- Resistance to heat, freezing, antibiotics or irradiation
- Their ability to alter otherwise unfavorable substances allowing for the outgrowths of pathogenic bacteria.
- It may cause off odors, off flavors and discoloration of food surfaces.
- Enumeration of yeasts and moulds in foods generally provide sufficient information for routine control procedures.

Reagents and Materials

1. Potato Dextrose Agar Media
2. Distilled water
3. pH meter
4. Autoclave
5. Laminar flow
6. Glasswares
7. Weighing balance

Media: Potato Dextrose Agar (Acidified)

Ingredient	*Quantity*
Infusion from white potatoes	200.0 ml
Dextrose	20.0 g
Agar	15.0 g
Distilled water	1.0 litre
Autoclave 15 minutes at 121°C	

Procedure

Media Preparation

- Suspend 39 g of commercial dehydrated ingredients in distilled water.

- ☆ Heat mixture to boiling to dissolve the ingredients.
- ☆ Distribute into tubes or flasks and autoclave 15 minutes at 121°C (15 lbs pressure).
- ☆ When used as plating medium for yeast and moulds melt in following steam or boiling water, cool and acidify to pH 3.5 with sterile 10 per cent tartaric acid solution.
- ☆ Mix thoroughly and pour into plates.
- ☆ To preserve solidifying properties of agar do not heat medium after the addition of tartaric acid.

Selective Action

Yeast and moulds need acidic pH (3.5) to grow and under the conditions unfavorable for bacterial growth. In Potato Dextrose Agar, the high concentration of starch and sugar at low pH 3.5, inhibit bacterial growth.

Procedure

- ☆ Preparation of sample and decimal dilution of appropriate concentration is made as described earlier.
- ☆ Prepare pour plates with 15 to 20 ml agar tempered to 45 °C ±1°C.
- ☆ Allow the Petri dishes to solidify, invert the plates and incubate at 20 to 25°C for 5 days.
- ☆ If excessive mould growth develops on the plates, count first after 3 days and then again after 5 days.
- ☆ Counting plates from the reverse side is sometimes helpful if they are overgrown with moulds.

Precautions

- ☆ Acidify the sterile and tempered medium with predetermined quantity of tartaric acid solution immediately before pouring plates.
- ☆ Do not reheat medium once acid has been added.

13. Coagulase Positive Staphylococci Determination

Food poisoning by enterotoxic strains of *Staphylococcus aureus* is often traced to meat products. It would therefore, be an obvious advantage if meats and meat product ingredients could be monitored for toxigenic strains of *S. aureus* and their presence either totally eliminated or held to very low level. It is important to detect coagulase positive Staphylococci, on the theory that coagulase and enterotoxin production are closely related.

Media: Trypticase soy Tryptose Broth

Ingredient	*Quantity*
Tryptose	10.4 g
Trypticase or Tryptone	8.5 g
Sodium chloride	5.0 g
Soy tone	1.5 g
Dextrose	1.8 g
Dipotassium phosphate	1.25 g
Yeast extract	3.0 g
Distilled water	1.0 liter
pH	7.2±0.2
Autoclave 15 minutes at 121°C	

Baird-Parker Medium

Ingredient	*Quantity*
Tryptone	10.0 g
Beef extract	5.0 g
Yeast extract	1.0 g
Sodium pyruvate	10.0 g
Glycine	12.0 g
Lithium chloride 6 H_20	5.0 g
Agar	20.0 g
pH	7.2±0.2
Autoclave 15 minutes at 121°C	

Brain Heart Infusion Agar

Ingredient	Quantity
Calf Brain (infusion)	200.0 g
Beef Heart (infusion)	250.0 g
Proteose peptone	10.0 g
Dextrose	2.0 g
Sodium chloride	5.0 g
Dipotassium phosphate	2.5 g
Distilled Water	1.0 litre
pH	7.4±0.2
Autoclave 15 minutes at 121°C	

Procedure

- ☆ Weigh aseptically 10 g of the meat sample into a sterile container or blended jar.
- ☆ Add 90 ml of sterile 0.85 per cent saline solution. Mix or blend. This is a 1:10 dilution.
- ☆ Prepare subsequent dilutions as described earlier.
- ☆ Enrichment procedure can be used by inoculation to Trypticase Soy broth having 10 percent sodium chloride concentration.
- ☆ Incubate at 35°C for 48±2 h.
- ☆ Streak on Baird-Parker agar plates using 3 mm loop.
- ☆ Incubate at 35°C for 48 h.
- ☆ Pick suspect colonies and inoculate a 13 × 100 mm tube containing 0.2 ml BHI broth
- ☆ Incubate the tubes at 35°C for 18-24 h.
- ☆ Add 0.5 ml reconstituted plasma (rabbit) with EDTA to the BHI tube. Mix thoroughly.
- ☆ Incubate at 35°C and examine at hourly intervals from 1 to 6 hrs for clot formation. Report as coagulase positive Staphylococci only those reactions that have a 4+ clot formations outline.

14. Salmonella Detection

The detection of salmonellae is based on four steps. Theses are (1) Pre-enrichment in a good growth medium (2) Enrichment in a medium less inhibitory for salmonellae than other bacteria (3) Plating on selective indicator agar and (4) Isolation from (3) and confirmation by biochemical and serological reactions. Since enrichment techniques are involved, quantitative estimates can be made only by inoculating the enrichment medium in an MPN sequence. The use of a pre-enrichment medium (step 1) is especially useful in examining a food product where contaminating salmonellae have been injured by heating, drying or similar processing steps (meat by products are a case in point). For fresh meat, it can usually be omitted with no sacrifice in the sensitivity of the assay. The following general instruction for the isolation and identification of salmonellae are adapted from the Bulletin, salmonellae in foods and feeds by Galton, Morris and Martin published by the National Communicable Disease Center US Public Health Service, 1968 and gives a useful outline of the method to be followed with meat and meat products.

Brilliant Green Agar (BGA)

Ingredient	*Quantity*
Yeast extract	3.0 g
Proteose peptone No. 3 or polypeptone	10.0 g
Sodium chloride	5.0 g
Lactose	10.0 g
Sucrose	10.0 g
Phenol Red	0.08 g
Brilliant Green 0.25 per cent solution	5.0 ml
Agar	20.0 g
Distilled Water	1.0 litre
pH	6.9±0.2
Autoclave 15 minutes at 121°C	

Dissolve ingredients by boiling in distilled water. Autoclave 121°C for 12 min. Cool to 45 to 50°C and pour 20 ml portions into 15X100 mm Petri dishes. Let dry for about 2 hours with covers partially open, then close plates. Final pH 6.9±0.2.

Procedure

- ☆ Weigh 100g of meat sample and mix it with 900 ml of sterile lactose broth in a blender jar.
- ☆ Transfer in a sterile container and incubate at 35°C for 24±2 h.
- ☆ After mixing the incubated lactose broth transfer 1 ml of it into 10 ml of selenite cystine broth and also into tetrathionate broth. Loosen the cap and incubate at 35° C for 24 ±2 hrs.
- ☆ After mixing the incubated broths, sterile xylose lysine desoxycholate agar and bismuth sulfite agar. Incubate at 35°C for 24 ±2 h.
- ☆ Examine for typical *Salmonella* colonies.

XLD Agar

Pink or red colonies with or without black centers.

BGA

Pink and red colonies.

BS Agar

Brown, gray or black colonies.

- ☆ If after incubation no growth occurred on any of the agar plates or if definite non-salmonella colonies appear, the test is concluded and sample reported as negative for Salmonella.
- ☆ Otherwise pick suspect colonies and inoculate triple sugar iron agar tubes and lysine iron agar tubes by streaking the slant and stabbing the butt. Incubate at 35° C for 24 ±2 h.
- ☆ Presumptive positive *Salmonella* in LIA are alkaline (purple throughout the tube. Hydrogen sulphide (black) may or may not be present.
- ☆ On TSI, presumptive positive cultures have alkaline (red) slants and acids (yellow) butts with or without hydrogen sulphide (black).
- ☆ Take presumptive positive TSI or LIA agar cultures and perform biochemical tests.

Urease Test

Subculture in 1.5 to 3.0 ml of urea broth and incubate at 35° C for 24 ±2 h. The typical *Salmonella* are urease negative offering no change in the orange color of the medium.

1. *Phenol Red Dulcitol broth*: Inoculate broth with presumptive positive culture and incubate at 35° C for 24 ± 2 h.
2. Gas and acid formation indicates a positive reaction (a yellow color).
3. *Phenol Red Lactose broth*: Inoculate broth with presumptive positive culture and incubate at 35°C for 24 ±2h. Most salmonellae give a negative test *i.e.* no gas formation and an alkaline reaction
4. *Phenol Red Sucrose Broth*: Same as Phenol Red Lactose broth
5. *Malonate broth*: Inoculate the broth with presumptive positive color and incubate at 35°C for 48±2 h. Most of *Salmonella* give a negative test without changing the original green color of the broth.

The presumptive cultures may be sent to specific laboratories for confirmation of *Salmonella* by polyvalent flagellar (H) screening test and polyvalent somatic (O) test.

15. Enumeration of Lactic Acid Bacteria

These are Gram +ve nonsporulating cocci or rods dividing in one plane only (with the exception of Pediococci), catalase–ve, usually nonmotile, obligate fermenters producing mainly lactic acid and sometimes also volatile acids and carbon dioxide. They are subdivided into the genera *Streptococcus, Leuconostoc, Pediococcus* and *Lactobacillus*.

Media Employed

1. Plate Count agar counting bromocresol purple (PCA-BCP)
2. Lactic Agar
3. MRS Agar/broth (Man-Rogosa and Sharp Agar) MRS Agar is used to support the growth of various lactobacilli particularly from dairy origin.
4. RMW Agar (Rogosa SL Agar) for cultivation of oral and faecal lactobacilli

5. Eugon Agar: to support surface growth of lactic acid bacteria.
6. APT Agar: For cultivation and enumeration of heterofermentative lactic acid bacteria of discolored cured meat products.
7. Acetate Agar
8. Separation of Lactic Streptococci

Reddy's Differential Broth for Separating Lactic Streptococci

Ingredient	*Quantity*
Tryptone	5.0 g
Yeast extract	5.0 g
Dipotassium phosphate	1.0 g
Arginine	5.0 g
Sodium citrate	20.0 g
Bromocresol purple	0.02 g
Distilled Water	1.0 litre

Media Preparation

- ☆ Suspend ingredients in approximately 800 ml distilled water, add 35 ml of 11 per cent reconstituted skim milk and bring volume to one litre.
- ☆ Steam the medium 15 minutes, cool to 25°C and adjusts pH to 6.2, dispense 7 ml quantities into 10×126 mm screw cap tubes containing Durham fermentation tubes.
- ☆ Sterilize at 121 °C for 15 min and final pH is adjusted to 6.2±0.2.
- ☆ Do not open the door until the temperature has dropped below 75 °C.

The common organisms used in the dairy products are *S. lactis, S. cremoris, S. lactis var. diacetilactis* which can be separated by biochemical tests–Arginine hydrolysis and test for diacetyl and acetoin.

Reddy *et al.*, 1971 developed a differential broth medium for differentiating these species. Individual colonies developing on agar plates were inoculated. After inoculation, closed test tube caps tightly to prevent escape of liberated NH_3 and CO_2. Incubate the tubes at 30°C for 24-72 h and observe indicator color reactions.

S. cremoris produces deep yellow color in the broth (acid). *S. lactis* initially turn the broth yellow (acid) but later the violet hue returns due to liberation of NH_3 from arginine. *S. lactis var. diacetilactis* yields a violet color and produces copious amount of CO_2 within 48 h from citrate in the fermentation tubes.

Reddy's Differential Medium for Enumeration of Lactic Streptococci

Ingredient	*Quantity*
Tryptone	5.0 g
Yeast extract	5.0 g
Casamino acids	2.5 g
Dipotassium phosphate	1.25 g
Calcium citrate	10.0 g
L-Arginine hydrochloride	5.0 g
Carboxymethyl cellulose	15.0 g
Agar	15 g

Procedure

Media Preparation

- Dissolve separately 10 g of calcium citrate and 15 g of Carboxy Methyl Cellulose by heating with continuous stirring in 500 ml distilled water.
- The other ingredients are separately dissolved and mixed with above solution.
- Adjust pH 5.6±0.2 with 6N HCl after heating. Sterilize at 121°C for 15 min.
- Just before pouring plates, add 5 ml sterile reconstituted nonfat milk (11 per cent solids), 10 ml of sterile 3 per cent calcium carbonate and 2 ml of sterile 0.1 per cent bromocresol purple in distilled water to 100 ml of sterile agar media melted and tempered to 55°C.

- ☆ Pour the plates and dry them for 18-24 h at 37°C in an incubator.

(*i*) For Differential Enumeration of Lactic Streptococci

- ☆ Spread 0.1 ml of the culture was on the surface of the above agar medium.
- ☆ Incubate plates in candle oats jar at 32°C and examine plates after 36h and after 6 days of incubation.
- ☆ On examining the plates, after 36-40 h, count all colonies and separately the yellow *S. cremoris* colonies.
- ☆ Further incubate for additional 4 days and expose the plates for 1 h to atmospheric air.
- ☆ First determine the total count and then count all colonies showing zones of clearing of the turbid suspension of calcium citrate (*Streptococcus lactis var. diacetilactis) S. lactis* colonies are obtained by subtracting *S. cremoris* colonies and *S. lactis var. diacetilactis* from total colonies.

Catalase Test

Grow the culture in a Petri plate and flood plates with 3.5 per cent H_2O_2 solution. Observe for bubble formation using a hand lens or with wide field binocular microscope. Colonies exhibiting no evidence of gas formation are catalase negative. Transfer a loopful of colonies to a slide. Mix with 2 to 5 per cent H_2O_2 solution and observe as above. Catalase negative is no evidence of gas formation.

(*ii*) Determination of Acid Producing Micro-organisms in Food (Meat)

Micro-organisms and its Significance in Food

Genera *Streptococcus, Leuconostoc, Pediococcus* and *Lactobacilus* produce mainly lactic acid and sometimes-volatile acids and CO_2. Many spore forming species belonging to genera *Bacillus* and *Clostridium* are important acid producers and cause quality deterioration and subsequent spoilage of foods particularly canned foods. Most enteric bacteria *e.g. Escherichia, Enterobacter, Salmonella and Shigella* are able to carry out a mixed acid type or butylenes glycol fermentation and these are human pathogens mostly responsible for food poisoning and dysentery.

Media

All-purpose media with Tween 80 (A.P.T. agar) with bromocresol purple (BCP).

Media Composition

☆ APT Agar (all purpose Medium with Tween 80)

APT agar is prepared by adding 15.0 g of agar to the basal formula below. Bring to a boil to melt the agar. Dispense and autoclave as below:

☆ APT agar + BCP

Prepare agar as indicated. Add 2 ml of bromocresol purple, 1.6 per cent in alcohol per litter of medium prior to dispensing.

APT Broth

Ingredient	*Quantity*
Trypticase, peptone or Tryptone	10.0 g
Yeast extract	7.5 g
Dipotassium phosphate	5.0 g
Sodium chloride	5.0 g
Sodium citrate	5.0 g
Sodium carbonate	1.25 g
Thiamine	0.0001 g
Dextrose	10.0 g
Tween 80 (polyoxyethylene sorbitan monooleate)	0.2 g
Magnesium Sulphate	0.8 g
Magnesium chloride	0.14 g
Ferrous sulfate	0.04 g
Distilled water	1.0 l

Dissolve ingredients in distilled water and dispense into flasks or bottles. Sterilize for 15 minutes at 121°C. longer sterilization is not recommended. The final pH should be 6.7±0.2.

Selective Action

Bromocresol purple is used as an indicator, which has a pH

range of 5.4 (yellow) to 7.0 (purple). BCP is used to enumerate different acid producing types of bacteria present in food products. Acid produced and accumulating around colonies will change the color of the indicator in the medium and identify those colonies to be counted.

APT agar was developed for the cultivation and enumeration of heterofermentative lactic acid bacteria of discolored cured meat products.

Micro-organisms vary in thiamine requirement. *Lactobacillus* spp. and other heterofermentive lactic acid bacteria require about 10-µg/litre, whereas *Lactobacilus viridescens* require at least 100 µg/litre for their growth. Lack of catalase production, high salt tolerance, ability to change the color of meat, ability to grow at low temperature, high acid tolerance, production of acetic acid and lactic acid is characteristics of these organisms. High nutrient contents also make APT media useful for detecting fermentive organisms.

Reagents and Materials

- ☆ Common equipments and glasswares available in microbiological laboratory.

Procedure

- ☆ Decimal dilutions of 10^{-3} and 10^{-4} was made.
- ☆ 1 ml of each dilution was put into each Petri dish used.
- ☆ Approximately 15 to 20 ml APT agar media was poured into the plates aseptically.
- ☆ Plates were mixed uniformly and incubated at 37 °C for 48 h.
- ☆ The colonies with characteristics color *i.e.*, yellow to purple were enumerated and interpreted.

16. Determination of Halophilic Micro-organisms in Food (Meat)

Microorganisms and its Significance in Foods

The slightly halophilic microorganisms (2 to 5 per cent salt concentration) belonging to the genera *Pseudmonas*, *Moraxella*, *Acinetobacter* and *Flavobacterium* contribute to the spoilage of marine fish and shellfish. Most of the moderately halophilic bacteria involved in the spoilage of salted foods (5 to 20 per cent NaCl) are Gram +ve spp of the *Bacillus, Micrococcus* and *Achromobacter*. The extremely halophilic micro-organisms (>20 per cent to 30 per cent NaCl) are principally species of genera *Halobacterium and Halococcus* and the halo tolerant microorganisms (>5 per cent NaCl as well as no NaCl) include Gram +ve bacteria like *Micrococcus* and *Bacillus* and *Staphylococcus aureus, Clostridium perfringens* and *Clostridium botulinum* causing food poisoning outbreaks involving salted foods. These organisms cause spoilage of marine fish, shell fish and crustacean, brined and cured meat products (ham,bacon, prepared meat and sausages), salted fish etc. A yeast known as *Debaryomyces* can bring spoilage changes in salted meat and fish.

Reagents and Materials

- ☆ Same as before.

Media: Sea Water Agar (SWA)

Ingredient	*Quantity*
Yeast extract	5.0 g
Peptone	5.0 g
Beef extract	3.0 g
Agar	15.0 g
Sea water (synthetic)	1.0 l

- ☆ Dissolve ingredients in the synthetic seawater and sterilize at 121°C for 15 min. Final pH should be 7.2±0.2.

Phosphate Buffered Saline (PBS) pH 7.6, 0.01M

Concentrated (10 x) stock solution (pH is not 7.6 at this point)

Disodium phosphate anhydrous, reagent grade: 12.35 g

Monosodium phosphate. H_2O (reagent grade): 1.80 g

Sodium chloride, (reagent grade): 85.0 g

Distilled water to a final volume: 1.0 litre

Working solution: PBS, pH 7.6, 0.01 M

Conc. Stock solution: 100 ml

Distilled water q.s.: 1litre

Selective Growth

The growth of the types of halophilic organisms is dependent upon the appropriate salt (NaCl) concentration.

Procedure

- ☆ Prepare dilutions by blending 50 g samples with 450 ml diluent (phosphate buffer with salt) with added NaCl equivalent to salt concentration of food sample.
- ☆ Prepare pour plates using seawater agar (SWA) media. If food sample is not refrigerated, use 1 ml aliquot for pour plates and incubate plates for 4 days at 25°C.
- ☆ For refrigerated foods such as bacon, plate 0.1 ml aliquots with the spread plate technique.
- ☆ Incubate plates 10 days at 7°C. Report results on counts per gram.

Observations

The presence of pink or red colonies is indicative of extreme halophiles.

17. Determination of Proteolytic Micro-organisms in Meat

Microorganisms and its Significance in Foods

Proteolytic species are common among the genera *Bacillus, Clostridium, Pseudomonas* and *Proteus.* Micro-organisms that carry out protein hydrolysis and acid fermentation are called acid proteolytic for example *Streptococcus faecalis var liquefactions* and *Micrococcus caseolyticus.*

Protein hydrolysis by micro-organisms in foods may produce a variety of odour and flavour defects. Some of the common psychrotrophic bacteria are strongly proteolytic and cause undesirable changes in dairy, meat, poultry and seafood products particularly when high populations are reached after extended refrigerated storage. In some foods the level of proteolytic microorganisms may be of value to project refrigerated storage life and to assess processing methods.

Reagents and Materials

- Same as in determination of psychrotrophs.

Media: Skim Milk agar

Composition

Standard Methods Agar or Plate Count agar: 1.0 litre

Reconstituted skim milk (10 per cent solids): 100 ml

Melt standard methods agar or Plate count agar. Cool to 50°C and add 100 ml sterile skim milk, mix well and pour into Petri dishes.

Selective Growth

The hydrolysis of casein occur by the proteolytics. These organisms convert casein into soluble nitrogenous compounds resulting in a clear zone surrounding the colonies of bacteria.

Procedure

- The meat samples are prepared and appropriate decimal dilutions are made as describe earlier.
- Pour plates are prepared using the skim milk agar media and 1 ml of the appropriate dilution with in each Petri dish.
- The prepared Petri dishes are incubated at 21°C for 72h or the incubation conditions for the total count may be used.

18. Determination of Lipolytic Micro-organisms in Meat

Microorganisms and its Significance in Foods

The genera *Pseudomonas, Achromobacter* and *Staphyloccus* among the bacteria, *Rhizopus, Geotrichum, Aspergillus* and *Penicillium* among the moulds and the yeast genera *Candida, Rhodotorula* and *Hansenula* contain many lipolytic species causing lipolytic degradation of foods. These organisms cause hydrolytic and oxidative lipolysis. Lipolytic and oxidative changes in meats and other fat containing foods are often associated with quality loss or spoilage under some storage conditions (*e.g.* high humidity).

Principle

Degradation of glycerol tributyrate (tributyrene) is indicated by brightening zones around the lipolytic colonies. This test compares favorably with butter fat test showing the capability of microorganisms of hydrolyzing fat.

Reagents and Materials

- ☆ Same as previous experiment. Preparation of sample and dilution is made as before.

Media: Tributyrene Agar

Ingredient	*Quantity*
Peptone	0.5
Yeast extract	0.3
Agar	1.5

- ☆ Suspend in 100 ml of distilled water.
- ☆ Adjust pH 7.5±0.1. Sterilize for 15 min at 121°C.
- ☆ Cool to about 80°C and add 1g of glycerol tributyrate already heated to 80°C.
- ☆ Swirl or mix it properly. The medium should exhibit uniform turbidity after cooling.

Procedure

- ☆ Serial dilutions 10^{-3} and 10^{-4} are made taking meat sample as described in earlier experiments.
- ☆ 1 ml of each dilution *i.e.* 10^{-3} and 10^{-4} is pippetted into the sterile Petri dishes in duplicates near the flame.
- ☆ Approximately 15-20 ml of sterilized tributyrine agar was poured into each Petri dish by carefully opening the Petri dish near the flame.
- ☆ The TBA medium is thoroughly mixed, solidified and Petri dishes are kept inside an incubator at a temperature maintained at 37°C for 3 days.
- ☆ Enumeration of colonies surrounded with bright zone was done and interpreted

19. Determination of Sulphide Spoilage Anaerobic Spore Formers

Microorganisms and significance in food: Sulphide spoilage may occur in non-acid canned foods. Spoiled product has a blackened appearance due to reaction between sulphide and iron of the container. The causative organism is *Clostridium nigrificance*. The cans involved show no evidence of swelling, but upon opening the cans, an odor of hydrogen sulphide is evident. The cause of sulphide spoilage is the combination of high spore numbers and holding of finished product at elevated temperatures.

Sulphide spoilage spores shall be present in not more than 2 (40 per cent) of the 5 samples tested and in any one sample to the extent of not more than 5 spores per 10g. This would be equivalent to 2 colonies in the inoculated tubes.

Reagents and Materials

- ✰ Same as before and anaerobic jars

Media

Sulfite Agar (For the detection of Thermophilic Anaerobes producing H_2S)

Ingredient	*Quantity*
Tryptone	10.0 g
Sodium Sulfite (anhydrous)	1.0 g
Agar	20.0 g
Distilled water	1.0 l

- ✰ Dissolve ingredients in distilled water, dispense into tubes in about 15 ml amount and into each tube place an iron nail or a small clean strip of iron nail or a small clean strip of iron.
- ✰ No adjustment of reaction is necessary.
- ✰ Autoclave 20 minutes at 121°C.
- ✰ Tubes should be used within a week after making.
- ✰ As an alternate for the iron strip or nail, 10 ml of a 5 per cent solution of iron citrate may be substituted in the sulfite medium formula. It is necessary to heat the citrate solution to completely dissolve the ferric citrate scales or pearls.
- ✰ Alternatively Tryptose sulfite Neomycin Agar (T.S.N. Agar) of Hi-Media pH 7.2±0.2 was used after rehydration @ 40 g/litre and sterilized.

Procedure

- ☆ 10 g of meat sample with 90 ml sterile distilled water was homogenized to make 1:10 dilution (10^{-1}).
- ☆ Subsequently 10^{-2} and 10^{-3} dilutions of the above mixture were made by decimal dilution method.
- ☆ 1 ml of 10^{-2} and 10^{-3} each dilution was pippetted into Petri dishes in duplicate aseptically.
- ☆ About 15 to 20 ml of melted T.S.N. agar media was poured into each petridish and mixed thoroughly, then kept for cooling and solidification of the media inside the Petri dishes.
- ☆ The Petri dishes containing the inoculums were kept inside an anaerobic jar, which was then transferred into an incubator maintained at 50°C and kept for incubation at this temperature for 24-36 hrs.
- ☆ After incubation was over, the microbial colonies (jet black spherical colonies) are counted, and interpreted.

20. Determination of Psychrotrophic Micro-organisms in Food

The microorganisms, which grow in foods at refrigerated temperatures, are usually called psychrophillic. Now more commonly the term 'psychrotrophic' is used. Species of *Pseudomonas, Achromobacter, Flavobacterium* and *Alcaligenes* are often included in this group of bacteria and *Geotrichum* and *Botrytis* are the molds able to grow in refrigerated foods. These organisms cause slime formation and a variety of off flavour, change in color, flavour and texture of meat as well as physical defects in foods. Their growth is temperature dependent. Raw foods, non-sterile heat processed foods, when kept in refrigerated temperature for extended period are spoiled by psychrotrophs. Presence of psychrotrophs (Bacillus and Clostridium) in pasteurized food implies post-processing contamination. Frozen foods (sometimes reprocessed) for retail sale as refrigerated food are thawed and show a large no. of psychrotrophs. Some microbes may grow slowly at –5 to –12 °C. Quality losses of foods occur due to prolonged storage.

Reagents and Materials

1. Pestle and mortar
2. Test tubes, pipettes
3. Petri dishes
4. 0-1 per cent peptone water for dilution
5. BOD incubator
6. Water bath etc.

Media: Plate Count Agar (Standard Methods Agar)

Ingredient	*Quantity*
Glucose	1.0 g
Yeast extract	2.5 g
Tryptone	5.0 g
Agar	15.0 g
Distilled water	1.0 l

- ☆ Dissolve the ingredients in distilled water by boiling and adjust to pH 7.1±0.1.
- ☆ Dispense into tubes or flasks and autoclave at 121°C for 15 minutes.
- ☆ Final pH should be 7.0±0.1.
- ☆ Alternatively readymade plate count agar of above composition can be used after rehydration @ 23.5 g/1000 ml distilled water.

Procedure

- ☆ Decimal dilutions of the sample (refrigerated meat) at 10^{-2} and 10^{-3} are used by pour plate method.
- ☆ The Petri dish incubated at 5°C for 7-10 days.

21. Isolation of Diferent Micro-organisms of Public Health Significance

(*i*) *Yersinia enterocolitica*

- ☆ Food samples are prepared by homogenizing 10 g in 90 ml buffered peptone water (BPW) containing 1 per cent peptone.

- These broths were then incubated at 4± 1°C for 2-3 weeks in refrigerator.
- At the end of the incubation period, samples were spread on Mc Conkey- Tween 80 Agar (MT) and held at 25°C for 24-48 h.
- MT agar was prepared by incorporating 10 g of Tween 80 and 0.2g of $Cacl_2$ per liter of Mc Conkey Agar (Oxoid, CM7).
- Suspect *Y. enterocolitica* colonies (2.0 mm diameter, translucent pink, wrinkled appearance and zone of sheen caused by hydrolysis of "Tween 80") were further investigated and sub cultured to Brain Heart infusion Agar (Oxoid CM 375) in order to prepare pure cultures and incubated at 27°C for 24 h.
- Then colonies were streaked on the Mc Conkey Agar (27°C, 48-72h), Kliger's iron Agar (KSI, Oxoid CM 33) (27°C48-72h), Christensen Urea Agar (27°C, 48-72h) and inoculated in Glucose broth (25°C, 24-48h) (degradation of glucose) for biochemical tests.
- Motility test in Brain Heart infusion Broth (BHI, Oxoid CM 225) (25°C, 24h) methyl red test in Methyl Red Voges-Proskauer (MR-VP, Oxoid CM 43) medium (25°C, 48-72h) were also carried out (Harrigan 1998).
- The reference strain of *Y. enterocolitica* (134, 0:3) was obtained from Pasteur Inst. Paris and included in all identification tests.

ii) Listeria monocytogenes

- Enrichment procedures are often required for this organism. Weigh 10g of food sample and homogenize with 90ml of enrichment broth (University of Vermont broth, UVM, or Fraser broth or modified UVM)
- Food samples are prepared by inoculating in selective or non-selective broths (nutrient broths).
- These broths are then incubated at 4°C for 8 weeks up to 12 weeks in refrigerator.
- Sub cultures are made at weekly interval and at the end of the incubation period on to blood agar with addition of antibiotic to prevent growth of Gram-negative

contaminants or selective agar like *Listeria* selective agar and held at 37°C for 24-48 h.

- Suspect *L. monocytogenes* colonies (0.5 to 2.0 mm diameter, transparent colonies with smooth borders appear on blood agar with narrow zones of beta haemolysis.
- *L. monocytogenes* can hydrolyze aesculin, which results in black zones around colonies due to formation of black, iron phenolic compounds.
- This organism shows characteristic tumbling motility when tested by hanging drop method.

(iii) *Bacillus cereus*

- Enrichment procedures are used in non selective media such as Nutrient broth or selected media as trypticase–soy-polymyxin broth
- Food samples are inoculated in these broths and incubated at 37°C for 18h. If Nutrient broth is used for enrichment, then a heat shock of 70°C for 15 min is recommended before incubation as it reduces the chances of growth of contaminants.
- Inoculums from these broths is sub cultured on selective agar *i.e.* mannitol egg yolk polymyxin agar (MEYP) or Polymyxin pyruvate egg yolk mannitol bromothymol blue agar (PEMBA)
- These are incubated at 37°C for 24 hours.
- Suspected *B. cereus* colonies (gray, white flat or slightly raised with opaque zone of turbidity) are identified on the basis of colonial morphological, and biochemical tests.

(iv) *Clostridium botulinum*

Meat has seldom been implicated in food poisoning by *Clostridium botulinum* in United States, although it frequently was in some European countries. This is no doubt due to differences in the characteristics of meat products made (salt, pH etc) and the general use of refrigeration in United States. *Clostridium botulinum* is a widely distributed soil organism of very common occurrence, and the quality control supervisor should be aware of its proclivities. It will grow where there is an absence of atmosphere oxygen, a neutral pH and an organic substrate. Salt and Salt nitrite combinations are inhibitory to the organisms but must be present in higher concentrations than usual for most products to be entirely effective. Heat treatments, as in processing, will destroy the vegetative organism but its spores are quite resistant and only heating under pressure, or for a long period as in canning, will destroy them. Low pH is highly inhibitory to *Clostridium botulinum* and is one of the best safeguards of many meat products

Enrichment procedures are often required for isolation. Robertson cooked meat medium (RCM) or reinforced Clostridial meat medium and trypticase-peptone-glucose-yeast extract broth (TPGY) with or without trypsin may be used.

- ☆ Dissolved oxygen from media is removed by steaming for 10-15 min followed by immediate cooling.
- ☆ Weigh 1-2g of food sample and inoculate two tubes of RCM and TPGY with samples such that it is introduced slowly beneath the surface of the broth.
- ☆ These broths are then incubated at 35°C for RCM and 26°C for TPGY.
- ☆ Tubes are examined for turbidity and gas production after 5 days otherwise examine incubate for further 10 days.
- ☆ In 1-2 ml of culture add equal amount of filter-sterilized ethanol, mix and incubate at room temperature for 1h. Heat shock may be given to the culture at 80°C for 10-15 min.
- ☆ Treated culture is then streaked on to blood agar or anaerobic egg yolk agar and plates are incubated anaerobically at 35°C for 48h.
- ☆ Suspect *C. botulinum* colonies (raised or flat, smooth or rough with irregular edges) appear on blood agar while on egg yolk medium precipitate under the growth is also visible.
- ☆ Select 10 colonies each and transfer into TPGY broth for non-proteolytic *C. botulinum* and RCM for proteolytic toxin types and incubate at 35°C for RCM and 26°C for TPGY.
- ☆ Cultures are streaked on to egg yolk agar plates in duplicate and on e plate is incubated anaerobically and other under

aerobic conditions at 35°C. Pure culture is indicated by no growth in aerobic plate and typical colonies in anaerobic plate.

Enterotoxigenicity

- Pure culture broths are taken from TPGY and centrifuged at 1200g for 10 min. The supernatant is divided into three parts, 1st is untreated, 2nd is treated with 0.2 ml of 10 per cent trypsin and 3rd is heat treated at 100°C for 10 min.
- 1st and 2nd part is diluted in the ratio of 1:2, 1:10 and 1: 100 with gel-phosphate buffer.
- Inject 0.5 ml of original and diluted fluids into pairs of mice intraperitoneally.
- Start observing mice 24h after injection till 48 h for the symptoms of botulism that includes flaccid paralysis, difficult breathing followed by death. No death of mice injected with heat treated fluid
- Calculate MLD50/ml and test further dilutions of toxin preparations if required

(v) Clostridium perfringens

Clostridium perfringens has often been found to be causative organism in food poisoning caused by meat products particularly cooked meat products. Growing use of convenience, ready-to-eat meat dishes will undoubtedly heighten the interest in *Clostridium perfringens.* The quality control laboratory, which includes a strong bacteriological capability, ought to prepare itself to monitor *Clostridium perfringens* in raw materials and products.

Enrichment procedures are often required for isolation. Thioglycollate broth, Robertson cooked meat medium (RCM) or lactose sulphate broth may be used.

Dissolved oxygen from media is removed by steaming for 10-15 min followed by immediate cooling.

- Weigh 1-2g of food sample and inoculate two tubes of thioglycollate broth with samples such that it is introduced slowly beneath the surface of the broth.
- These broths are then incubated at 44°C for 24-48h.

- 0.1ml of inoculum is added in the sterile Petri plate and pour molten selective plating medium {Tryptose sulphate cycloserine agar (TSC) or suphate polymyxin sulphadiazine (SPS) or Shahid Ferguson perfringens (SFP) agar} and incubated at 37°C for 24 h under anaerobic conditions.
- Suspect *C. botulinum* colonies (Black in SPS, black colonies surrounded by opaque zone in SFP) are picked up and processed further for identification.

Enterotoxigenicity

- Pure colonies are inoculated in Duncan and strong medium and incubated at 37°C for 24h for production of toxins.
- Methods used to detect toxins are mouse lethality, rabbit ileal loop test, vascular permeability reaction and cell culture using vero cells
- Calculate MLD50/ml and test further dilutions of toxin preparations if required

(*vi*) *Escherichia coli*

- Enrichment procedures are often required for isolation of *E. coli*.
- Food sample is inoculated in enrichment broth (Enterobacteriaceae enrichment broth, Mc Conkey broth) in the ration of 1:10 after mixing in a stomacher and incubated at 37°C for 24h.
- Culture are streaked on to selective media {Mc Conkey lactose agar (MLA), Eosin methylene blue agar (EMB) and 4-methyl eumbelliferyl β-D glucuronide (MUG) sorbitol as selective medium for O157:H7 *E. coli*} and incubated at 37°C for 24h.
- Suspect *E. coli* colonies (small round and pink colored on MLA, round with typical metallic sheen on EMB, pale colonies of O157:H7 do not give fluorescence under UV light where as colonies of all other serotypes appear pink and gives out fluorescence under UV light)

- Characteristic colonies are inoculated on nutrient agar and are further processed for identification on the basis of cultural, morphological and biochemical characteristics.
- Finally cultures should be sent for serotyping.

(*vii*) Salmonella

- Pre-enrichment and enrichment procedures are often required for isolation of *Salmonella*.
- 25 g of food sample is inoculated in pre- enrichment broth (Buffered peptone water) and incubated at 37°C for 16-18h.
- Sufficient inoculum is transferred to enrichment broth {Selenite cystine broth (SCB), Tetrathionate broth (TTB) and Rapport-Vassiliadis (RV)} and incubated at 42°C for 16-20h. Second enrichment if required may be done by inoculating loopful of growth to fresh 10 ml enrichment broth.
- Culture are streaked on to selective media {Brilliant Green Agar (BGA), Bismuth sulphite agar (BSA)} and incubated at 37°C for 24h. Second streaking from enrichment broth is also recommended from enrichment broth.
- Suspect *Salmonella* colonies (pink surrounded by red medium on BGA, Black centered, bright edged colonies surrounded by a precipitate showing metallic luster giving rabbit or fish eye appearance on BSA) are transferred to triple sugar iron agar tubes and incubated at 37°C fro 24h to see alkaline slant (pink) and acidic butt (yellow). Colonies showing agglutination with Salmonella poly O sera are also further processed for identification on the basis of cultural, morphological and biochemical characteristics.
- Finally cultures should be sent for serotyping.

(*viii*) *Staphylococcus aureus*

Method–I

- 10 g of food sample is inoculated in enrichment broth (staphylococcal enrichment broth or trypticase soy broth with 10 per cent sodium chloride and 1 per cent sodium pyruvate) and incubated at 37°C for 24h.

- ☆ Cultures are streaked on to selective media {Baired Parker agar (BPA) and incubated at 37°C for 24-48h.
- ☆ Suspect *Staphylococcus aureus* colonies (1-1.5mm diameter, shiny, jet black) are transferred to 0.2 ml brain heart infusion (BHI) broth tubes and incubated at 37°C fro 18-24h. Isolates are identified on the basis of cultural, morphological and biochemical characteristics. Coagulase test can be carried out using rabbit plasma

Method–II

- ☆ Take 63g Baird Parker Agar base was suspended in 950 ml distilled water and boiled to dissolve the medium completely.
- ☆ Sterilized by autoclaving at 15 lbs pressure (121°C) for 15 min.
- ☆ Cooled to 50°C, added aseptically 50 ml concentrated egg yolk emulsion and 3 ml sterile 3.5 per cent potassium tellurite solution and mixed well before pouring.
- ☆ Final pH of the medium was 7.0±0.2. 0.2 ml of suitable dilutions in duplicate was spread onto the surface of the medium.
- ☆ The petridishes were incubated at 35°C for 48 h.
- ☆ The number of intensely dark, shiny, regularly shaped colonies surrounded by clear haloes were counted and expressed as $\log_{10}$ CFU/g.

Enterotoxigenicity

- ☆ Pure colonies of *Staphylococcus aureus* are inoculated in brain heart infusion (BHI) broth.
- ☆ Cell suspension from broth is collected from supernatant fluid after centrifugation and preserved with 1:10, 000 merthiolate solution.
- ☆ The toxins were detected by using the optimal sentivity plate method using antisera.

22. Estimation of Total Anaerobes Count

- ☆ Suspended 58g anaerobic agar in 1 litre distilled water.

- ☆ Boiled to dissolve the medium completely and sterilized by autoclaving at 15 lbs pressure (121°C) for 15 min.
- ☆ Final pH of the medium was 7.2±0.2.
- ☆ Take 0.2 ml of suitable dilutions in duplicate and inoculate into the medium by surface spread method.
- ☆ The petridishes were put into the anaerobic jar and incubated at 35°C for 48 h.
- ☆ White colonies on the surface of the medium were counted and expressed as $\log_{10}$CFU/g.

23. Microbial Challenge Testing (MCT)

MCT is an established technique with in the food industry. It aims to simulate what can happen to a product during processing, distribution and subsequent handling, following inoculation with relevant microorganisms in relevant numbers. The test product has to be processed under appropriate and well controlled manner and then held under a range of controlled conditions that relates to those used by consumers. MCT has several areas of application such as determination of product safety, establishment of shelf life, formulating products in terms of intrinsic control factors such as pH, a_w etc have been identified world over.

Procedure

MCT can be conducted with any known spoilage and pathogenic micro-organism as per the method described by Notermans *et al*. (1993).

- ☆ First check the initial count of to be tested microorganism into the food/meat product using the specified selective media by pour plate method.
- ☆ Take a chunk of 1g of food/meat product under sterile conditions.
- ☆ Inoculate the chunk with the specific microbial inoculum and kept at the specified conditions of storage or serving for a specified period.
- ☆ Take out the chunk and estimate the microbial count of the specified micro-organism on the specified selective media by pour plate method.
- ☆ The counts were represented as cfu/g of each bacteria against which challenge testing was done

Explanation

Initially prepare the appropriate dilution of the meat product and enumerate the E.coli count using Eosin Methylene Blue (EMB) Media using pour plate method and incubation at 35°C for 48 h. Assume it is 1×10^{-1}cfu/g.

Then take a portion of meat product under sterile condition and spread the inoculum of known concentration of *E. coli* (say 1×10^{-6}cfu/g). Keep this portion of the product under the prescribed storage condition (may be refrigerated) for a period of 24 hr. Then take 1g of sample from the portion of the product and prepare decimal dilution. Pour plate 1ml of appropriate dilution into MB media and after incubation count the colonies. The increase or decrease in number of count than inoculated dose will reflect the potential of meat sample to resist the growth of *E. coli* and its shelf life. It can be further utilized for deriving the prediction equations for microbial life.

24. Methods for Rapid Detection of Micro-organisms in Meat and Meat Products

A number of indirect methods have been developed to rapidly assess the microbiological quality of meat and meat products. These methods are generally based on the physical or biochemical changes in the food samples and analysis of microbial cell components and metabolites present. These methods are grouped into two categories.

1. Procedure taking less then one hour called Real Time Procedure.
2. Retrospective assays requiring incubation up to 72 hrs.

(*i*) Real Time Procedure

(*a*) Bioluminescent/Adenosine Triphosphate (ATP) Assay

Bacterial cells have ATP in relatively constant amounts and can be used as an index to estimate bacterial numbers. The test is based on firefly bioluminescent reaction in which lucifern/luciferase reacts with ATP. However, to ensure selective measurement of microbial ATP it is essential to remove non-microbial ATP. This is generally eliminated by enzymatic degradation using various enzymes like apyrase 3. Another way is to separate non-microbial ATP from bacterial population by centrifugation followed by cation exchange to remove food particles and supernatant by membrane filtration.

(*b*) Direct Epifluorescent Technique

Involves use of slide processor and a fluorometer. Samples of enrichment culture are filtered, stained and rinsed and total fluorescence intensity is noted.

(*c*) Limulus Lysate Assay

Widely used for detection of bacterial endotoxins. The test is based on activation of a Ca^{+2}-dependent clotting enzyme by bacterial endotoxins, which are lipopolysaccharides. This follows cleavage of a coagulogen into peptide chains by the activated enzyme and formation of a stable gel protein matrix. This sensitivity of the technique is up to picrogram levels.

(*ii*) Retrospectives Methods

(*a*) Microcalorimetry Measurements

It was noted that heat liberated during spoilage of a canned food products could serve as a technique for identifying bacterial spoilage. In this procedure heat output from incubated samples is correlated with bacterial counts. The exothermic heat production rates (HPRS) are measured in a Bio Activity monitor and correlated to $\log_{10}$ colony forming units (cfu) per ml or g.

(*b*) Radioimmune Assay

In general RIA utilizes a molecule containing radio active atom. The procedure involves four components, an unlabelled antigen, a labelled antigen, a binder or antibody and the immuno absorbent material. The reactants are incubated together in glass or polystyrene tubes. The radio activity of one or each fraction is determined and a disc response curve is plotted from the standard data. A modification of RIA is immuno radiometric assay (IRMA) which utilizes radiolabelled antibody in place of radiolabelled antigen as in RIA.

(c) Enzyme Immune Assays

EIA can be classified into 2 systems: Competitive binding and non-competitive binding. The first one includes enzyme linked immunosorbent assay (ELISA). The antibodies are attached to a solid phase and then incubated with a fixed amount of enzyme-linked antigen.

Free antigens are then separated followed by addition of substrate to determine the activity of bound enzyme. In non-competitive assays, antigen is linked to a solid phase and bound with antibodies. This follows addition of second enzyme-linked antibody.

Chapter 8
Molecular Techniques in Meat Industry

Nucleic acid based techniques popularly known as molecular techniques are commonly used in meat industry for two major purposes one Meat Species Identification and second for the microbial quality evaluation. These techniques involve DNA analysis. It includes Polymerase Chain Reaction (PCR), Restriction Fragment Polymerase Chain Reaction (RFLP), Random amplified polymorphic DNA (RAPD) finger printing, DNA hybridization and DNA sequencing. In an animal cell, DNA is present in nucleus as well as mitochondria. The mitochondrial DNA is one of the highly conserved sequence in different species of animals. The high homogeneity of mitochondrial DNA is exploited to prepare primers that can amplify DNA of several other species including microbes.

Meat Species Identification

Principle

The meat and blood samples from cattle, buffalo, sheep, goat, pig and chicken are collected from various authentic sources. The high quality genomic DNA is extracted from meat and meat products processed under different manufacturing conditions. The DNA can also be isolated from blood samples. After checking the quality, purity and concentration of DNA, the PCR is optimized for amplification of target DNA. The PCR products obtained were analyzed on horizontal submarine agarose gel electrophoresis to determine size of DNA fragments. These DNA fragments are further confirmed by characterization with restriction enzyme analysis. Subsequently, the PCR assay is tested for its consistency with samples from different breeds. The PCR assay is cross tested for its specificity with DNA of other meat species. The PCR assay is also tested for its efficiency to amplify for DNA extracted from cooked and autoclaved meat and meat emulsion. Finally, the sensitivity of standardized PCR assay is tested in admixed meat and meat products at 5 per cent, 1 per cent and 0.1 per cent level.

Method

Step I: Meat and Blood Sample Collections

1. The fresh meat samples of buffalo, sheep, goat, pig and chicken should be collected.
2. The samples should be stored at –20°C till further analysis.
3. The blood is collected in sterile 15 ml polypropylene tube containing 0.5 ml of 0.5 M ethylene diamine tetra acetic acid (EDTA) solution, which acts as an anticoagulant.
4. The collected blood samples also preserved at –20°C till DNA isolations.

Reagents and Materials

The chemicals and reagents used should be of molecular biology grade of high purity and standard.

(*a*) Thermocylcer
(*b*) Electrophoresis Assembly
(*c*) Micro-centrifuge
(*d*) Micropipettes
(*e*) Gel Documentation System
(*f*) Table Top Centrifuge
(*g*) Spectrophotometer
(*h*) Cyclo Mixer
(*i*) Digital Dry bath

(*j*) Ice Flaking Machine
(*k*) Electronic Balance
(*l*) Freezer
(*m*) Refrigerator
(*n*) Meat Mincer
(*o*) Bowl Chopper

Oligonucleotide Primer Pairs

- ✰ Published primer pairs based on the mitochondrial gene sequences of cattle, buffalo, sheep, goat, pig, and chicken should be used.
- ✰ The published DNA sequences of the different species for designing of primers are retrieved from the NCBI GenBank.
- ✰ The Species-specific primer pairs are designed from conserved sequences of different regions/genes of mitochondrial genome using softwares like Gene-Tool and DNASTAR.
- ✰ The primer pairs designed are synthesized from METABION INTERNATIONAL (HYSEL), GERMANY.

Step II: DNA Extraction of Genomic DNA from Muscle tissue/ Blood

Method–I

High quality genomic DNA can be extracted by using standard readymade kits (using *DNeasy* Blood and Tissue Kit (Qiagen, Germany) as per the manufacturer's instructions). All the centrifugation steps are performed at room temperature (15-25°C). Ethanol was added to buffers AW1 and AW2 before DNA extraction. The thermo-mixer is preheated at 56°C. DNA of cooked and autoclaved meat and meat products is also extracted by the same kit. The protocols followed are as given below.

DNA Extraction from Muscle Tissues

1. The meat tissue is cut into the small pieces, 10.25 mg of tissue is placed in 1.5 ml microcentrifuge tube and 180 µl buffer ATL is added.

2. Proteinase K (20 µL) is added and mixture is vortexed briefly followed by incubation for 1-3 hr at 56°C or until the tissue is completely lysed.
3. Two hundred microlitre each of Buffer AL and ethanol are added separately to the mixture. The mixture is vortexed for 15 sec before and after adding the buffer and ethanol.
4. This mixture is put into the *DNeasy* Mini spin column into a 2 ml collection tube. The tube is centrifuged at 8000 rpm for 1 min.
5. The column is placed in a new 2 ml collection tube and 500 µL buffer AW1 is added. The tube is centrifuged at 8000 rpm for 1 min and flow through is discarded from collection tube.
6. The column is once again placed in a new 2 ml collection tube and 500 µl buffer AW2 is added. The tube is centrifuged at 14000 rpm for 3 min. the flow through is discarded from the collection tube.
7. Finally, the column is transferred into a new autoclaved 1.5 ml or 2 ml micro-centrifuge tube and 200 µl buffer AE is added for elution of DNA from spin column. This is incubated for 1 min at room temperature and centrifuged at 8000 rpm for 1 min. this step is repeated once for maximum yield.

DNA Isolation from Blood

1. In a 1.5 or 2 ml micro-centrifuge tube, 20 µl Proteinase K is placed and then 50-100 µl mammalian (5-10 µl avian) anti-coagulated blood is added. The volume is adjusted to 220 µl with PBS.
2. The buffer AL is added and mixed by vortexing. The tube is incubated at 56°C for 10 min. this is followed by addition of 200 µl ethanol (96-100 per cent) and mixing by vortexing.
3. This mixture is put into the *DNeasy* Mini column placed into a 2 ml collection tube. The tube is centrifuged at 800 rpm for 1 min and flow through is discarded from collection tube.
4. The column is placed in a new 2 ml collection tube and 500 µl bufferAW1 is added. The tube is centrifuged at 8000

rpm for 1 min and flow through is discarded from collection tube.

5. The column is once again placed into a new 2 ml collection tube and 500 µl buffer AW2 is added. The tube is centrifuged at 14000 rpm for 3 min and flow through is discarded from collection tube.
6. Finally, the column is transferred into a new autoclave 1.5 ml or 2 ml micro centrifuge tube and 200 µl buffer AE is added for elution of DNA from spin column. This is incubated for 1 min at room temperature and centrifuged at 8000 rpm for 1 min this step is repeated once for maximum yield.

Method–II

1. The stored blood samples were thawed and made uniform solution.
2. The tubes are filled with chilled RBC lysis buffer (1X), mixed end to end, incubated in ice for 10 mins and centrifuged at 3000-4000 rpm for 15-20 mins at room temperature.
3. The reddish tinged supernatant containing plasma and lysed RBC can be discarded by simple inversion of tubes or by pipetting.
4. Two volumes of chilled RBC lysis buffer are added, the tube ends are tapped to disperse the pellet of WBC or RBC (if left) and centrifuged at 3000-4000 rpm for 15-20 mins.
5. The black tarry colored supernatant containing lysed RBC is discarded by simple inversion.
6. The above steps can be repeated till the WBC pellet appeared nearly white free from RBC.
7. Add 3ml of DNA extraction buffer (@3ml/10 ml of blood) in the tube, shake it to disperse WBC pellet and keep it in incubator (25-30°C) for 30 min to ensure that the buffy coat is completely suspeneded so that the cells are accessible to SDS and proteinase K.
8. Add 200µl of 10 per cent SDS @200µl/10 ml of blood and gently invert the tubes to mix the content.

9. Add 25µl of proteinase K (@ 25µl of 20mg proteinase K/ ml of td H_2O for 10 ml blood) in two times *i.e.* half the requirement is added and mixed gently and kept for 3-4 hours in water bath at 50°C. After 3-4 hours the remaining amount of proteinase K is added and tubes are incubated at 50°C overnight.
10. After that, transfer the contents into a clean, sterile, autoclaved polypropylene tubes and add equal amount of equilibrated phenol (Tris saturated phenol pH>7.8), mixed by inverting gently for 15 mins till a light coffee colored uniform solution is formed. Centrifuge it at 3000-4000 rpm for 15-20 mins.
11. Transfer the upper aqueous phase containing DNA into fresh 15 ml clean, sterile, autoclaved polypropylene tubes.
12. Similar extractions (as above) can also be done once with equal volume of phenol: chloroform: isoamyl alcohol (25:24:1) and with chloroform: isoamyl alcohol (24:1).
13. Add 3 M sodium acetate (pH 5.2) @ 100µl/iml to the aqueous solution and gently mix it.
14. Add two volumes of chilled ethanol/ room temperature isopropanol and mix it gently and allowed to precipitate DNA.
15. Transfer DNA along with 500µl isopropanol with the help of micro pipette and centrifuge it 10000 rpm for 10 mins.
16. Discard the supernatant
17. Wash the DNA pellet twice in 500 µl of 70 per cent ethanol and again centrifuge it at 10000 rpm for 10 mins at room temperature.
18. Carefully air dries the DNA pellet on blotting paper to remove the traces of ethanol.
19. Add 200 µl of TE buffer and keep in water bath at 60°C for 2hrs to inactivate DNAse and other enzymes.
20. Store it till analysis at –20°C.

Quality of DNA

Horizontal submarine agarose gel electrophoresis is used to check the quality of genomic DNA using 0.8 per cent (w / v) agarose gel. At first, the gel-casting tray is prepared by sealing its ends with adhesive tape and the comb is arranged over it in such a way that a gap of 0.5 mm remained between the tips of the comb teeth and floor of the casting tray, so that the wells get completely sealed by agarose.

Subsequently, 0.8 per cent agarose (w / v) suspension in IX TBE buffer is made and heated on an electric heater until the agarose is completely melted to give a clear transparent solution. After cooling to 60°C, ethidium bromide (10 mg/ml) @ 5 μl per 100 ml of agarose solution is added and mixed gently. The agarose is poured into a leveled casting tray and the gel is made to about 4 mm in thickness. The agarose is allowed to set at 4°C temperature till it solidified and subsequently the comb and adhesive tapes are gently removed.

For loading the sample, 5 μl diluted DNA is taken and after mixing it with 2 μl of 6 X gel loading dye, it is loaded into the well of agarose gel which is submerged in the electrophoresis tank containing 1X TBE buffer. A DNA marker is also run in one of the wells. Electrophoresis is performed at 80V for 80-90 min. After electrophoresis the gel is visualized under UV transilluminator and documented by gel documentation system.

Purity of DNA

The purity of DNA is checked using UV Spectrophotometer (BECKMAN DU 640, USA). 6 μl of DNA of each sample is diluted in 294 μl of triple glass distilled water and Spectrophotometric readings at OD_{260} and OD_{280} ratio are taken against 300 μl triple distilled water as a blank. The DNA samples of $OD_{260:280}$ ratios between 1.7 to 1.9 are considered good and are used for PCR amplification.

Quantification of Genomic DNA

In general two common methods are used to measure the amount of nucleic acid in an extracted sample

Spectrophotometric Measurement

- ✰ The optical density of nucleic acid sample is measured at 260 and 280 nm wavelength in an ultra violet spectrophotometer.
- ✰ The standard curve of optical density of known different concentration of DNA is prepared and concentration of unknown is calculated. However in sample mixed with RNA or oligos the standard measurement of known

concentration of RNA or oligo is also measured as a reference.

$1\ OD_{260}$ of double stranded DNA = 50 µg/ml
$1\ OD_{260}$ of single stranded DNA = 33 µg/ml
$1\ OD_{260}$ of single stranded RNA = 40 µg/ml
$1\ OD_{260}$ of oligos = 20 µg/ml

☆ Pure preparation of DNA and RNA have OD_{260}/OD_{280} values of 1.8 and 2.0 respectively.

Concentration of DNA

For estimating the concentration of DNA, following formula was employed

$$\text{DNA concentration } (\mu g/\mu l) = \frac{OD_{260} \times (\text{Dilution factor}) \times 50}{1000}$$

(1 OD value at 260 nm is equivalent to 50 ng dsDNA/µl)

Optimization of Polymerase Chain Reaction (PCR) Assay

Reagents

Stock and Working Solution of each dNTP

The stock solution of 100 mM of dNTP is prepared and subsequently working solution is prepared by reconstituting the dNTP's. 10 µl of each d NTP is taken and volume is made upto 100 µl *i.e.* 10 µl dATP + 10 µl of dGTP + 10 µldTTP + 60 µl of tdH_20. The effective concentration of each dNTP becomes 10 mM in the mix.

Primer Reconstitution

Stock of X nm is dissolved in X µl to get 1000 pm/ µl. From this stock solution of 1000pm/ µl, working solution of 20pm/ µl is prepared by diluting the 1 µl of stock to the volume of 50 µl. store it –20°C for further use.

PCR Buffers

☆ 25 mM $MgCl_2$,
☆ 10 X PCR buffer with KCL

Method

- Prepare the reaction mixture in a 500 µl PCR tube to a total volume of 50 µl containing 5 µl of 10X PCR buffer (with KCI), 3 µl of $MgC1_2$ (25 mM), 1-2 µl of dNTP mix (10-20 mM each), 0.25-0.50 µl of Taq DNA polymerase (1-2.5 units), 1-2 µl each of forward and reverse primer (10-20 pmol), 1 µl of DNA template (20-30 ng) and nuclease free water (to make up the reaction volume to 50 µl). This mster mix is further used for the analysis.
- The PCR tube containing reaction mixture is flash spun on a table top micro-centrifuge to get the reactants at the bottom.
- Take all the ingredients using filter tips to avoid any cross-contamination.
- Every-time negative control (without template DNA) is always put to make sure that there is no contamination in PCR system.
- The cycling conditions are:
  - Initial denaturation at 94°C for 2 min
  - 30-35 cycles (45 cycles for double PCR at 0.1 per cent level of adulteration)
  - Denaturation at 94°C for 0.5 min,
  - Annealing at optimized temperatures and extension at 72°C for 5 min,
  - Pause at 4°C.
- Finally, the PCR product is kept at –20°C for further use.

Electrophoresis of PCR Products

Horizontal submarine agarose gel electrophoresis is used for analysis of PCR products using 1-2 per cent (w / v) agarose gel. The agarose gel is prepared as mentioned in section 3.6.1. 5-10 µl of PCR products are mixed with 1-2 µl of 6 X gel loading dye and loaded into the well of agarose gel which is submerged in the electrophoresis tank containing 1X TBE buffer. A 100 bp DNA ladder is also run in one of the wells. Electrophoresis is performed at 80V for 80-90 min. once the electrophoresis is over, the gel is visualized under UV transilluminator and documented under gel documentation system.

Sequencing of PCR Amplified DNA Fragments of Mitochondrial 16s rRNA Sequence

PCR product after checking the correctness of amplification should be checked for thesequecing at the designated laboratories.

PCR products were sequenced using ABI Prism 377 DNA sequence at DNA sequencing facility, University of Delhi, South Campus, New Delhi. The sequences obtained can be analyzed using Laser gene software (DNA-STAT). The comparison of sequence can be carried out by using Clustal V method with Megalign™ software package (DNA-STAR). The mitochondrial 16S rRNA gene sequences of cattle, buffalo, sheep, goat and pig can be sequenced and their sequence comparison can be made for meat species identifications.

Restriction Mapping of Sequences

The sequences obtained can be restriction mapped with the help of DNA-STAR software. Enzymes having restriction sites helpful in species identification are chosen by tabulation and comparison.

Characterization/Restriction Fragment Length Polymorphism of PCR Products

The results of species-specific PCR assay can be further confirmed by digestion of PCR products with various restriction enzymes. The PCR products of mitochondrial 16S rRNA gene amplified by designated primers can be subjected to restriction enzyme (RE) digestion with selected enzyme for that species. The procedure for RE digestion of PCR products can be carried as per the details given below:

- ☆ Purify the PCR products by using PCR purification kit.
- ☆ The restriction digestion reaction can be assembled by adding the given reagents in following order: Nuclease free water, PCR product (10-20 µl), 10X enzyme buffer (EB) and restriction enzyme.
- ☆ Incubate the reaction mix overnight at 37°C in water bath.
- ☆ Stop the digestion reaction by adding 6X loading dye.
- ☆ Keep the products at –20°C till electrophoresis.

- ☆ The digested products can be subjected to electrophoresis in 2.5 per cent agarose gel along with 100 bp ladder.
- ☆ Finally after electrophoresis, observe the gel for desired band pattern and documented by gel documentation system or photography.

Specificity of PCR Assay

The species-specific PCR assay can be cross tested with pooled DNA samples of all species referred for meat speciation with primer pair supposed to be specific for one species at optimized conditions. Observe the amplification pattern for single fragment in one species without any cross reaction with other species.

Sensitivity of PCR Assay

The sensitivity of species-specific PCR assay can also be tested in admixed meat and meat products under different processing conditions. Various combinations of meat mixtures can be prepared containing 5 per cent, 1 per cent and 0.1 per cent of meat to be detected in total 250 gm meat and meat emulsion. The sensitivity can also be checked by 10-fold serial dilution of DNA extracted from meat and meat products.

Coded Samples Analysis

The coded samples of meat and meat products containing varying proportions (5 per cent, 1 per cent and 0.1 per cent) of meat of unknown species can also be checked qualitatively presences or absence of meat species for authentication of the primers.

Microbiological Quality Evaluation

The meat and meat products can be evaluated for the presence or absence of any specific spoilage or pathogenic microorganism. It is quicker method to quantify the specific microorganism in the sample than the traditional method of enrichment and isolation on the selective media. The basic steps of the molecular techniques remain the same it differs in few regarding sample preparation and template preparation of the specific microorganism.

Microbial Template Preparation

The characteristic colonies of the strains of specific microorganism grown on the selective media are picked and grown on the enrichment media. Take 1 ml of the pure culture and centrifuge it at 10,000 rpm for 10 min. Discard the supernatant and dissolve the pellet in 500 µl of autoclaved milli-Q-water and then subjected to vigorous heating in boiling water for 10 min. Transfer the micro centrifuge tubes immediately in ice to snap chill. From this about 5 µl was used as template in PCR.

The procedural steps of Standardization of PCR and confirmation through agarose gel electrophoresis remain the same.

Chapter 9
Sensory Evaluation of Meat Products

Introduction

Sensory evaluation/analysis is the scientific way of judging meat and meat products by using human senses of sight, smell, taste, touch and hearing, tactile, temperature, pain etc. This is an important tool to assess the acceptability of a product to human palate. This evaluation is considered necessary for new product development, analysis of competitive products, product shelf life studies, market level consumer test, etc.

Difference Between Organoleptic Testing and Sensory Evaluation

In traditional organoleptic evaluation, it seldom complies with the principles of the science (full objectivity and reproducibility). Panel members evolve in organoleptic testing often feeling rather than analysing his impression; relying on experience, confusing quality tests with hedonic evaluations; and not checking on accuracy of his sensory abilities.

The modern sensory analysis requires a trained sensitivity and a reliable, precise and reproducible rating of small, often minute, differences in sensory perception. The panel members judge the quality of products by using his senses adopting exact methods of evaluation (*e.g.* difference test, descriptive test etc); participation in appropriate training courses and periodic selection tests; and working in a panel where test results are analysed statistically.

General Testing Conditions

1. Panel leader
2. Test subjects- sensitivity (normal olfactory and gustatory sensitivity), age, sex, smoking habits, health, additional points (taking drinks, spicy foods, sucking candies or chewing gums, use of cosmetics or perfumes etc), avoidance of disturbances during test (noise, off-odors etc), the number of test subjects (20-30)-hedonic scale (30-50).
3. Personnel
4. Sensory laboratory
5. Panel room
6. Time to test
7. Number of tests
8. Testing beakers
9. Samples
10. Testing technique
11. Fatigue and neutralization
12. Test sheets
13. Presentation, discussion and interpretation of results

Sensory Evaluation Methods

There are two methods for assessing sensory characteristics like appearance and color, flavour, tenderness, juiciness and overall acceptability of meat and meat products

Objective Methods

They involve use of instruments such as Instron, Warner Bratzler shear force, Flavour Analyzers, Optical nose etc. for above mentioned sensory characteristics. For example:

Texture Profile Analysis of Meat Foods By Texture Analyzer

Texture is the response of tactile (touch) senses to physical stimuli that result from contact between some part of body and the food. The tactile sense (touch) is the primary method for sensing texture but kinesthetics (sense of movement and position) and sometimes sight (degree of slump, rate of flow) and sound (associated with crispiness, crunchy and crackly texture) are also used to evaluate the texture. Texture is the primary consideration in determining the eating quality of fish, meat and their processed derivatives. In texture profile analysis several variable are measured under controlled conditions by using multiple measuring instruments like Texture Analyzer or Instron. In texture profile analysis, a bite size piece of food is pierced/compressed/ sheared one or two times in a reciprocating motion that imitates the action of human jaw and a number of parameters are extracted from the resulting force-time curve that correlate well with sensory properties of those parameters. A variety of probes and load cells along with appropriate software are available with the instrument to make it versatile for different types of analytical applications.

The rheological properties of emulsion or dough are also important to determine the textural and other sensory attributes of processed meat products. The textural properties of meat products are measured using a texturometer (TMS-PRO Texture Measurement System, Food Technology Corporation, USA or Texture Analyser, Stable Micro System, UK or Brookfield Texture Analyzer, UK).Texture Profile Analysis (TPA) can be performed as per the procedure outlined by Bourne (1978).

Materials

- ☆ Texture Analyzer
- ☆ Selection of right fixture/probe depends upon the type and condition of meat and type of analysis. It varies from fresh meat, meat emulsion dough (semi-solid), cooked product etc. *e.g.* Extrusion- extrusion cell/extrusion cone/ extrusion platen set, Shearing- Standard single blade shear cell, Multiple blade shear cell, Penetration/fracturability- single needle, multiple needle, compression cell etc.
- ☆ Selection of load cell also varies according to type of product and type of analysis. In general for the meat and meat products samples, load cell of 25-100kg is used.

Preparation of Sample

The samples selected for the estimation of texture analysis should be of uniform size. In general it is suggested that samples should be drawn from the frozen/ chilled carcasses/cut-up-part should be taken with the help of mechanical borer. The samples should be drawn from the same location of different carcasses for the comparison of results. The orientation of fibers within the cored muscle sample presented to test blade will directly affect the force required to shear them.

Cut a piece of food 1 cm^3 using a cutting die (appropriate size depending upon type of food material). Same size and shape should be used for replicates.

Allo Kramer Shear Press

This is estimated by using Kramer Shear Cell. Bundles of muscle fibers tightly grouped together. The muscle pieces are kept in such a manner that the orientation of mucle fiber should be at right angle to the blades of shear cell and 100kg load cell is used at a test speed of 25mm/ sec. The results reproducibility is improved when single orientation of fibers is used.

Illustration

Various tests can be conducted for the analysis of complete texture characteristics. However, in general texture profile analysis (TPA) of meat products is conducted using double compression cycle and different parameters are calculated from the graph (force vs. time / force vs. distance) framed during the process. The brief description is given below:

Performance of Test

Before proceeding for test, caliberate for force and probe as per procedure given in the manual of instrument. Force calibration is required to be done after change of load cell. Once it is calibrated, no calibration is required until load cell of another capacity is replaced.

Now enter the 'settings' into instrument as pretest requirements. The following settings are generally used in a TMS-PRO Texture Analyser (Food Technology Corporation, USA) model as mentioned below:

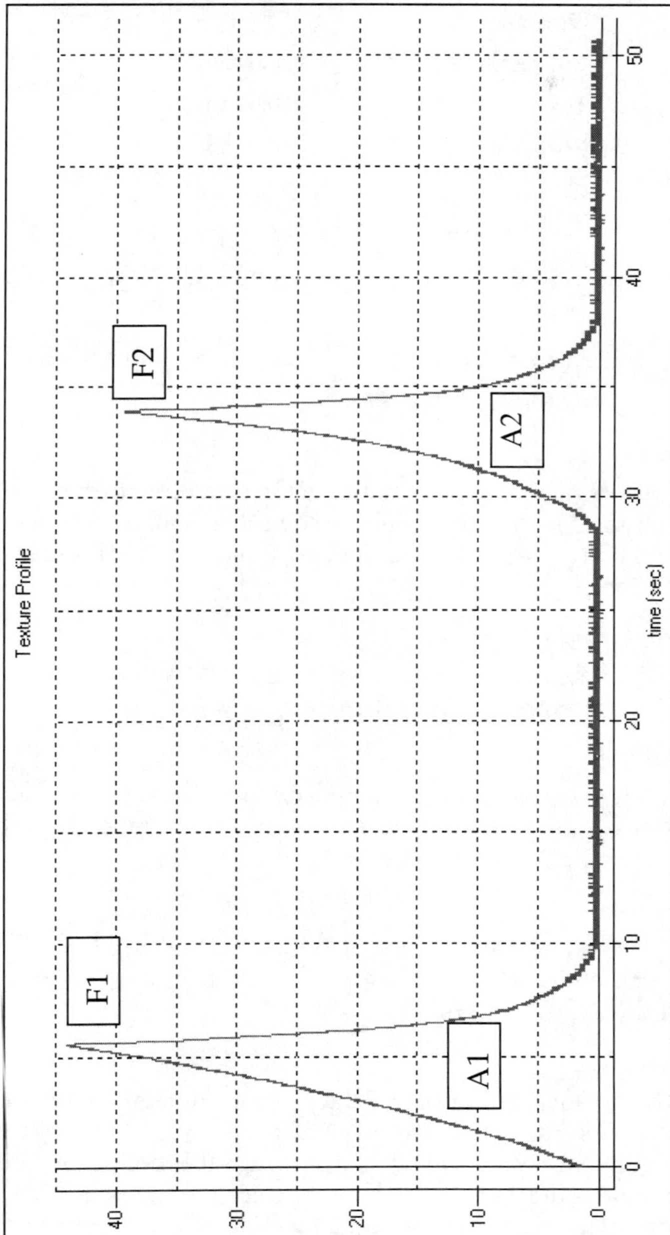
Texture Profile
F1
F2
A1
A2
Load N
0
10
20
30
40
time (sec)
0
10
20
30
40
50

Option	:	TPA 2
Pre-test speed	:	10.0 mm/s
Test-speed	:	10.0 mm/s
Post-test speed	:	10.0 mm/s
Compression 1	:	50% strain
Compression 2	:	25% strain
Trigger Type	:	Auto-5 g
Time	:	3 s
Data Acquisition rate	:	2000 pps (points per second)

Interpretation of Curve

Hardness

Sensory description is amount of force required to bite through sample. It is the height of the force peak (F_1) on the first compression cycle (first bite is defined as hardness). It is expressed in g/N (force).

Fracturability/Brittleness

It is define as force of the significant break in the curve on first bite. It is also expressed in g/N (force).

Cohesiveness

It is the degree to which sample deforms before shearing. It is defined as ratio of positive force areas (work done) under the second peak to first peak compression *i.e.* A_2/A_1.

Adhesiveness

It is the negative force area (work) of the first bite (A_3). It represents the work necessary to pull the compressing plunger (probe) away from sample *i.e.* N/s.

Springiness

The sensory description is the previewed degree with which the sample returned to the original height and thickness after pressing five times. It is the distance, which the food recovered its height during the time that elapsed between the end of the first bite and start of the second bite. Originally this is known as elasticity. It is calculated from the ratio of time from start (B) to mid second peak (C) to the time from start (A) to mid of first peak (D) *i.e.* T_2/T_1.

Gumminess: This is the degree to which sample sticks to the mouth. It is defined as the product of hardness and cohesiveness *i.e.* $F_1 X (A_2/ A_1)$. It is expressed in N.

Chewiness (work): The sensory description is number of chews required to prepare the sample for swallowing. It is defined as product of gumminess and springiness *i.e.* $F_1 X (A_2/ A_1) X (T_2/T_1.)$. It is expressed as N/s.

Resilience: It is the flexibility of the product. Higher the value of resilience, more the stability of the product. It is the ratio of the area of second half to the area of first half of the first peak of texture profile curve.

Objective Estimation of Meat Color

Hunter Lab Mini Scan XE

The instruments like color difference meters have developed which measure the surface color of food materials within few seconds. For the determination of color, the food surface should be smooth and free from surface characteristics like gloss, texture and pattern. The processed foods are placed into a glass sample cup whereas fresh fruits and vegetable can be directly exposed to the port of instrument.

Principle

The color is three-dimensional characteristic of appearance consisting of lightness (value), hue and chroma. When all radiant light is reflected from an opaque surface, it appears white, however, when light is completely absorbed by the surface, it appears black. When there is partial absorption of light over entire visible spectrum, the color of object is grey. This overall relation of reflectance to absorption without regard to specific wavelength is termed as lightness or value. It is the 'L' value measured by color difference meter.

Particular color (hue or color) appears when there is more reflection of incident light at a particular wavelength than other wavelengths. When there is more reflection at 400-500 nm, hue is blue (blue color) and when there is more reflection at 600-700 nm, the color is red. The amount of light reflected at a given wave length gives rise to intensity of color (chroma or strength or purity of color). The colors can be distinguished from one another by specifying these three visual attributes *i.e.* lightness, hue and chorma. The color perception is further limited by the light source, optical geometry of instrument and the angle at which light is reflected. Therefore, the instruments are provided with standard illuminants and color is viewed at fixed angle which is specified for the quality characteristic to be determined.

Hunter Lab Color Difference Meter directly measures the color in terms of 'L', 'a' and 'b' values and color measured on this scale can be converted into other color scales described by International Commission On Illumination (C.*I.E.*). Hunter 'L' values give lightness and vary from 100 for perfect white to zero for black, similar to that which eye evaluates it. Hunter 'a' values are measure of redness (positive or +ve) or blueness (negative or–ve). Together 'a' and 'b' may provide result for hue and chorma dimensions. Instrument software provides intercovertible scales. Zero value of 'a' and 'b' represent grey color.

Reagents and Materials

1. Hunter Lab Color Difference Meter (Model: Mini Scan Xe plus or Bench Top)
2. Sample Cup Holder Assembly.
3. Glass sample cup 2.5 inch (64 mm)

Working of Color Difference Meter

Mini Scan Xe plus portable instrument is available in sphere (d per cent 80) and 45 0/O 0 optical geometry Sphere (diffuse/8 0) geometry minimizes the surface and only measures differences due to sample color. A 45 0/O 0 geometry measures apparent color changes due to both the sample color and surface shine or texture. The letter perceives the appearance like human eye. The viewing area of instrument can be either small or large.

The instrument geometry is either 45 0 illumination with O 0 viewing or diffuse illumination with 8 0 viewing (d.80). The label on the bottom of the instrument identifies the model, viewing area and geometry of the instrument.

The standardization on a Mini Scan Xe Plus model 45 0/O0 geometry requires reading of black glass and white tile. Standardized before proceeding with sample measurement and then at least once after every eight hours. Large area view (25 mm) and small area view (7.6 mm) options are available. It is xenon-flash spectrophotometer. It has D 65 illuminant (*i.e.* it represents daylight with correlated temp 6500K).

Procedure

Preparation of Sample for Measurement

Food sample may be chopped or made into a paste or sheet or even whole samples may be used. Directionality can be minimized by averaging several measurements with rotation of sample between readings. Examination of standard deviation (SD) display with the average function may be used as a guide for selecting the appropriate number of readings.

Non-opaque samples must have a consistent backing. A white uncalibrated title is recommended. Translucent samples should be measured with a large port opening than illumination area to allow light to escape. Light trapped in the sample can distort the color.

When measuring, it is important to selected samples appropriately, use an established measurement method and handle all samples in a consistent manner. Choose samples randomly that are representative of whole lot. Prepare samples in exactly the same manner each time they are measured. Present the samples to the instrument in a repeatable manner. The results measured depend on the condition of the sample and their presentation. For any established procedure, make a checklist so that operators may simply check each step.

Measurements

Calibrate (standardize) the instrument using light trap/black glass and white tile provided with the instrument. Select the suitable parameters and color scale. Thereafter, fill the homogenized food sample in the glass sample cup and place it on the reflection port of the instrument and take three readings. There are six color scales with the Mini Scan Xe Plus Color Difference Meter. Now compare the color reading with those provided for standard products in literature. Alternatively the standards of color quality may be established by evaluating the color by sensory method and co-relating with the reading of the instrument. Another alternative may be the instrumental readings of the color charts available for different foods products and thereafter taking the color readings of these food samples to assess their color quality.

In this exercise students are advised to select the food products of different color quality (grades) and establish the color grades on different scales for different products.

Precautions

1. Give warm up time to the instrument as per instructions given in instrument manual.
2. Identify the geometry of the instrument before use.
3. Use black glass or light trap for standardization of instrument as per geometry of the instrument.
4. Restandardize instrument after every 8 hr of use.
5. Prepare and present sample in consistent way to instrument for determination.

Lovibond Tintometer Color Units

The fresh meat color of ground meat is compared using a Lovibond Tintometer (Model E, U.K.). Samples from three different places of meat mass were taken in the sample holder and secured against the viewing aperture. The sample color was matched by adjusting red (*a*) and yellow (*b*) units, while keeping the blue units fixed at 1.0. The corresponding color units were recorded. Hue and Chroma values are calculated using the following formulae (Little, 1975; Froahlich *et al.*, 1983):

$$\textit{Hue value} = (\tan^{-1})\, b/a$$

$$\textit{Chroma} = (a^{2} + b^{2}),$$

where, a = red unit, b = yellow unit

Subjective Methods

These methods include perception through human senses. It involves the specialized panel or consumer panel for testing the sensory attributes of meat and meat products. These tests are more authentic than objective methods.

Even though sophisticated and highly sensitive measuring instruments such as GC, mass spectrometers, NMR spectrometers, IR and UV spectrophotometers etc., are now available, the importance of the subjective methods of sensory analysis has grown rather than diminished. This is due to at the limit of instrumental sensitivity, where no signal appears; our biological detector may still perceive an odour, taste etc. Further, the instruments only analyse single compounds, whereas our senses give us a total impression of aroma, taste, temperature and tactile components.

Subjections in Sensory Evaluation

Sensory evaluation can be divided into the following five subjections-

1. Difference tests
2. Threshold and dilution tests
3. Ranking tests
4. Descriptive and rating tests
5. Hedonic tests

1. Difference Tests

Difference test implies to ask whether or not there is a difference between the samples. Additional questions concerning differences in intensity or type may be asked and are of special use for product testing and panel training and selection. There are three difference tests:

(*a*) The paired difference test

(*b*) The triangle test

(*c*) The duo-trio test

(a) Paired Difference Test

In this test, two samples are provided and the analysis is done for a single attribute. The trainee has to identify the sample having more intensity of the specified attribute.

Paired difference test is of two types

(*i*) *Direct paired difference test*: Several pairs are given to each panel member. Each pair consists of a control sample and an analytical sample. The question is posed which sample in each pair is sweeter, saltier, more sour etc.

(*ii*) *Second type paired difference test*: Pairs with same and different samples are presented. The question is posed whether the samples in each pair are identical or different.

Observation-Paired Difference test:

Samples	*Attribute*	*Interpretation*
A		
B		

(b) Triangle Test

The triangle test is the most widely used of all difference tests. In this test, three coded samples are provided to trainee and he has to identify which two samples are same or which one is different from other two.

There are two types of triangle tests

(*i*) *Simple triangle test*: In this test, three samples are presented, two are alike, one is different and question is posed which is the odd sample. For example: AKK KAK KKA and AAK AKA KAA

(*ii*) *Extended triangle test*: In this test there are additional tasks, may be defining degree of difference, and/or characterizing the difference. The triangle test is of equal importance in product testing and panel training and selection.

Observations

Samples	*Attribute*	*Interpretation*
A		
B		
C		

(c) Duo-trio Test

In this test, three samples are presented to trainee. One of them is reference sample and other two are coded samples. The trainee has to identify that among the coded samples, which one is similar or different to reference sample. The duo-trio test is so called because it is intermediate between the duo (paired) and the trio (triangle) test. The test samples contain one sample marked control (K) and one or several pairs of samples. Each pair contains one control sample and an analytical sample in random order. The question posed is " which is different sample in each pair?" It can also be asked, "Which sample in each pair is identical to control?"

K-KA or *K-AK*

KAK

AK

KA

AK

KA

Observations

Samples	*Attribute*	*Interpretation*
A	Reference sample	
B		
C		

2. Threshold and Dilution Tests

Threshold and dilution tests may be conducted to determine the sensitivity of the candidates to the four basic tests (sweet, sour, salty and bitter). They are also important in tests with aroma chemicals to determine the sensitivity of the biological detector in comparison to measuring instruments. There are four different thresholds:

(*a*) *Detection or stimulus threshold*: When a solution in a concentration series tastes different to water, even though the basic taste cannot be recognized, then the detection threshold reached.

(*b*) *Recognition threshold*: A concentration of a solution at which the basic taste is recognized correctly, is known as recognition threshold.

(*c*) *Terminal threshold*: Above a certain concentration, the basic tastes no longer be differentiated. This concentration is terminal threshold.

(*d*) *Different threshold*: It is defined as that concentration at which difference is recognized.

This test is used for training of the panel members for making them familiar to four basic tastes. The serial dilutions in ascending order of 0.01 per cent of all the four tastes (cane sugar-sweet; sodium chloride-salty; tartaric acid-sourness; caffeine-bitterness) are prepared. The trainee is given reference sample of each basic taste than they are provided with serial dilutions of each taste solution in ascending order until he is able to recognize the taste. The concentration of particular sample is considered as Threshold value of trainee for that specified taste.

Interpretation

Threshold value for an average good taste is 0.68 per cent cane sugar solution for sweetness, 0.019 per cent tartaric acid solution for sourness, 0.18 per cent sodium chloride for saltiness and 0.009 per cent caffeine solution for bitterness.

Precautions

- Rinse your mouth with potable water between successive samples.
- Never taste more than six samples continuously to avoid fatigue of taste.
- The ideal time for conduct sensory evaluation of meat and meat product is late morning or late afternoon.
- Never compel any one for sensory evaluation.
- Never perform the sensory evaluation full stomach.

3. Ranking Test

Ranking test is for the classification of the samples according to differences in one or several specific quality components in the series to be ranked. It can be performed

- In increasing intensity of property (sweetness, saltiness, aroma etc.)
- For quality
- For hedonic properties

This method has the advantage that more than two samples can be compared at the same time. But disadvantage can be inexact results in the case of small differences and large variations.

A series of test samples are provided in an ascending or descending order of intensity of specific characteristics. The objective is to select one or two best samples out of all tested samples. The trainee receives a random order of sample and is asked to rank the samples in order of increasing intensity.

Observations

Product Samples	*Rank*	*Interpretation*
A		
B		
C		
D		

4. Descriptive Sensory Analysis

(i) Flavour Profile Test

The best known method of descriptive sensory analysis is flavour profile. This technique provides a detailed, descriptive evaluation of the quantitative and qualitative attributes of a flavour complex. The profile method concentrates on the entire flavor of a product and the individual attributes of a flavor in relation to each other. The dimensions of flavour analysis by the profile method include: (1) perceptible aroma, taste, flavor and feeling factors (called 'character notes'), (2) degree of intensity of each factor, graded on the following scale:

O	=	Not present
X	=	Just recognizable or threshold
1 or +	=	Slight
2 or + +	=	Moderate
3 or + + +	=	Strong

(3) order in which these factors are perceived, (4) after taste (5) amplitude or overall impressions of aroma and flavor, graded on the following scale:

X	=	Very low
1	=	Low
2	=	Medium
3	=	High

Flavour profile test utilizes a panel of 4 to 6 selected judges who first examine and then discuss the product in a open session. Once agreement is reached on the description of the product, the panel leader summarizes the results in report form. Panelists are selected for training based on a series of screening tests including sensory acuity, interest, attitude and availability. The panel leader plays a key role in directing the conversation and providing a conclusion for the test. Panelists could be led to a conclusion without being aware that this had occurred. This test has to be carried out very thoroughly over a long period of time. Only repeated analysis (10, 15 or 20 times) with the same test product will provide uniform results by a whole panel.

(ii) Quantitative Descriptive Analysis (QDA) Method

In QDA method, approach is primarily behavioural in orientation with considerable emphasis on the use of replication as a basis for assessing the quality of the output. The QDA method is based on the following principles. The development of the method evolved from a member of major considerations to ensure that it would:

- ☆ Be responsive to all the sensory characteristics
- ☆ Be a multi product test
- ☆ Use a limited number of subjects (10-12)
- ☆ Use subjects who are qualified before participation
- ☆ Employ a language develop process from leader influence.
- ☆ Be quantitative
- ☆ Have a useful data analysis system
- ☆ Have a data processing capability

Application of Descriptive Analysis

- ☆ Monitor competition
- ☆ Storage testing

- ☆ Product development
- ☆ Quality control
- ☆ Instrument sensory relationships.

5. Hedonic Scaling

It is the subjective hedonic preference of the consumer or taste panel members and is rated depending upon their liking and disliking or acceptance-rejection attitude. In hedonic scaling, effective responses of like and dislikes are measured on a rating scale. It is also a descriptive scaling of the characteristics of the product. The hedonic scale rating reflects the attitude of a group of people toward certain foods under a given set of conditions. Most popular are 8-point or 9-point Hedonic scale, where 8 or 9 = like extremely and 1= dislike extremely. In this respect given below some score sheets:

Proforma for Organoelptic Evaluation of Meat Production

Name of Panel member : Expt. No.

Product Name : Date

Scoring Guide : 9-point Haedonic scale

Very desirable	9
Desirable	8
Moderately desirable	7
Slightly desirable	6
Neither desirable nor undesirable	5
Slightly undesirable	4
Moderately undesirable	3
Undesirable	2
Very undesirable	1

Wash your mouth with clean water after tasting each sample, please.

Remarks:

Signature:

Proforma for Organoleptic Evaluation of Meat Production

Name of Panel member : Expt. No.

Product Name : Date

Scoring Guide : 8-point descriptive scale

Sensory Attributes	*8-Point Scale for Descriptive Attributes of the Product*							
	8	*7*	*6*	*5*	*4*	*3*	*2*	*1*
Color/Appearance	Excellent	V.good	Good	Fair	Sli. Poor	Mod. Poor	Very Poor	Extr. Poor
Flavour	Extremely desirable	Very desirable	Mod. desirable	Slightly desirable	Slightly undesirable	Mod. undesirable	Very undesirable	Extr. undesirable
Texture	-do-	-do-	-do-	-do-	-do-	-do-	-do-	-do-
Juiciness	Extremely juicy	Very juicy	Mod. juicy	Slightly juicy	Slightly dry	Mod. dry	V.dry	Extr. dry
Overall Acceptability	Extremely Acceptable	Very much Acceptable	Mod. Acceptable	Slightly Acceptable	Slightly unaccep-table	Mod. unaccep-table	Very unaccep-table	Extr. Acceptable
Sample Code	Color/ Appearance	Flavour	Texture	Juiciness	Overall Acceptability			

Remarks: Signature:

Evaluation of Fresh Meat

1. Time Intensity Test

This may be dual attribute time intensity (DATI) or single attribute time intensity (SATI) method. In SATI method, two individual meat samples are required to evaluate juiciness and tenderness separately, while in DATI only one sample is required to assess both attributes concurrently. The panelists (8 member) are presented with DATI test on computers and one trained to move a mouse diagonally across a mouse pad to record both attributes simultaneously (toughness-vertical scale and juiciness–horizontal scale) The scales are labeled with appropriate descriptors juiciness: not juicy (0) and very juicy (60); force to chew; low force (0) and high force to chew (60).

2. Line Scale Profile Method

This method is used for evaluating cubes for tenderness, juiciness, beef flavour, chewiness, moisture absorption and time to chew. 12 panelists are trained to evaluate six attributes using a 10 cm line

Based on the published literatures it is found that sensory evaluation can be conducted using trained panel for mutton (species–related flavour) intensity on a 9-point scale (1 = extremely weak; 9 = extremely strong), warmed-over flavour intensity on a 5-point scale (1 = none; 5 = very pronounced) and for color score 5 pt. scale (1 = pale pink, 2 = pink, 3 = pinkish red 4 = bright red, 5= reddish brown) and odour score 5 pt scale (1 = very unpleasant, 5 = very pleasant) of ground buffalo meat and chevon. Ultrasonic spectral feature analysis was conducted for measuring beef sensory attributes.

Visual color score, 5 pt. scale (Sahoo,1995)

1= Pale pink; 2= Pink; 3 = Pinkish red; 4= Bright red; 5= Radish brown.

Kansas State University Beef Color Scale (Shivas *et al.*, 1984)

1 = Very bright red; 2 = bright red; 3= slight dark red or brown; 4 = dark red or brown; 5 = extremely dark red or brown.

Odour Score, 5 pt. Scale (Sahoo,1995)

1= Very unpleasant; 2= Moderately unpleasant; 3= Unpleasant; 4= Moderately pleasant; 5= Very pleasant

Evaluation of Processed Meat Products

The available published literatures indicated that the panelists recorded their perceptions of each sensory attribute of ground beef patties on score cards containing 15 cm lines with anchor words. Scores could range from 1 to 15 for each attribute. The anchors were flavour (weak to strong) texture (tough to tender), juiciness (dry to moist) and after taste (none to strong). Color measurements (L, a and b) of the surface of the beef were made with a Hunter lab Labscan spectrophotometer and those of buffalo meat can be done by Livibond tintometer. The objective estimation of meat color is described earlier. Cooked patties are evaluated for beef, bloody, sour, grassy, spoiled and metallic aromatic impression by a 6-member sensory panel using a 15-point attribute scale. Sensory quality of frozen chicken meat sausages incorporated with skin, gizzard, heart and yolk was evaluated on a 7-point hedonic scale. Descriptive sensory evaluation for hardness, springiness, cohesiveness, juiciness and wheat flavour of smoked sausages made with mechanically separated poultry meat and wheat protein, was conducted by a panel of 9 judges on a 15 cm line intensity scale. Sensory evaluation of steam cooked chicken steaks was carried out by a panel of members using a 8-point hedonic scale and goat ham by 9-point hedonic scale. A panel containing 20 trained subjects evaluated the tenderness of beef. Pork and chicken cuts (hot and freshly roasted) using time intensity (T1) technique.

The objective assessment is conducted for the color of meat (myoglobin, oxymyoglonbin and metmyoglobin) by spectrophotometer, surface color (L, a, b) by Hunter lab or Lovibond tintometer, for flavor of meat *i.e.*, odoriferous volatiles by gas chromatography, mass spectrometry and liquid chromatography, juiciness by measuring water holding capacity and tenderness by Warner-Bratzler shear device, Lee-Kramer shear press and universal testing machine.

Chapter 10

Testing of Packaging Films

1. Determination of Thickness of Packaging Films

Thickness can be defined as the perpendicular distance between the inner and outer surface of a packaging material. It has a profound influence on the physicochemical and strength properties of the packaging films, *e.g.* low density polyethylene films of 100 gauge and 150 gauge thickness have tensile strength of 3.7 and 4 kg/15 mm, elongation value of 300 and 250 per cent, bursting strength of 0.98 and 0.90 kg/cm2 and tearing resistance of 27 and 18g, respectively in machine direction. Water vapour and gas transmission rates of plastic films also decrease with increase in thickness *e.g.* high density polyethylene films of 200 gauge and 300 gauge thickness have WVTR of 1.9 and1.5g/m^2/24 hr at 38°C and 90 per cent RH.

Apparatus Required

- ☆ Thickness gauge tester/screw gauge/micrometer
- ☆ calulator

Procedure

- ☆ First check the least count of the thickness gauge meter/ screw gauge/micrometer.

- ✰ Correct any error if any in thickness gauge meter/screw gauge/micrometer
- ✰ Thickness of plastic films is measured on dial gauge or micrometer.
- ✰ Generally, ten readings are taken at different places in the film and their average is taken as a measure of thickness.
- ✰ High quality films do not show much variation in thickness measurement at different points.

Note:

- ✰ Thickness of plastic films is measured in gauge in British system, mil in American system, mm in metric system and micron (μm) in continental system. The following factor can be much help in their inter-conversion:

  100 gauge = 1 mil
  = 1/1000 inch
  = 0.001 inch
  = 0.025 mm
  = 25 micron (μm)

2. To Measure the Strength Properties of Packaging Films

(*a*) Tensile Strength

The strength of a packaging film determines its resistance to rupture when subjected to a pulling force. It is expressed in kg/15 mm width. High tensile strength is necessary for flexible packaging materials to hold heavy packages and when packages are formed in semi-automatic or automatic pouch forming and filling machines and also in operations such as coating-laminated or printing. Most of the films and laminates have a tensile strength of 3-6 kg/15 mm width.

Apparatus Required

- ✰ Tensile testing machine

Procedure

- ✰ Standardize the equipment.

- ☆ Measure the tensile strength in both machine and transverse direction.
- ☆ The tensile strength is less in transverse direction than in machine direction.
- ☆ All strength parameter in packaging films are taken in both the direction because the values differ.
- ☆ The molecular orientation of a packaging film in machine direction and transverse direction are very much different.

Note:

- ☆ Elongation property of a packaging film is measured on the tensile strength tester itself. It denotes the length to which the film can be stretched by using a pulling force. It is expressed in percentage.
- ☆ Food packages which are likely to experience drop during distribution require material of higher elongation value of 200-500 per cent, whereas some plastic films like LDPE show an elongation value of 300-350 per cent.

(*b*) Bursting Strength

Bursting strength is a measure of the pressure which a packaging material sheet can resist before bursting. It is expressed in kg/cm^2. The bursting strength values can be used as a rough guide to compete the general strength properties of packaging films. Most of the unsupported films and laminates have bursting strength values between 0.7 and 2.0 kg/cm^2.

It is measured by burst strength tester which operates on the uniform pressure distribution principle of hydraulics. The fluid expands under uniformly increasing pressure against a distensible rubber diaphragm and simultaneously into a pressure gauge.

Apparatus Required

- ☆ Burst strength tester

Procedure

- ☆ Clamp the test material properly into the machine and hydraulic pressure is applied.

- ☆ As the material bursts, the pressure drops suddenly, but the indicator remains static to indicate the exact pressure at which the bursting occurred.
- ☆ Note this pressure, it will represent the bursting strength of the material.

(*c*) Puncture Resistance

It gives a reliable indication of the protection by packaging materials and containers against puncture hazards in handling and transit. It is expressed in Oz/tear inch and is measured by puncture resistance tester.

(*d*) Tearing Resistance

This is the measurement of the resistance to the propagation of a slight tear in the flexible packaging material. It is expressed in gm. This property is important for packaging materials which are used to package sharp edged food products. Tearing strength depends on orientation of molecules in the packaging material. PE has a very high tear resistance values since high elongation is normally associated with high tear resistance.

3. Chemical Identification of Different Plastic Films

The packaging materials can be identified with the simple chemical procedure. It is step by step method ruling out one category of the material and others.

Reagents and Materials

1. Toulene
2. CCl_4
3. Ethyl acetate
4. Cyclohexane
5. Hydrochloric acid
6. Sulphuric acid 40 per cent
7. Test tubes
8. Bunsen burner
9. Holder

10. Scissors
11. Round bottom flask
12. Pipette

Procedure

- ☆ Cut one square inch sample of unknown film.
- ☆ Place the sample in a 50ml round bottom flask and add 15 ml of toluene solvent.
- ☆ The flask is filled with water cool reflux condenser to prevent solvent loss.
- ☆ If the materials get dissolved than it may be Polyethylene or polypropylene or polystyrene.
- ☆ To differentiate further take fresh sample and add ethyl acetate, if it get dissolved into it, it is polystyrene.
- ☆ If, it is insoluble then it may be polyethylene or polypropylene.
- ☆ The sample insoluble in toluene can be further treated with CCl_4. The rubber HCl or polycarbonates are soluble into it whereas polyster or nylon are not soluble into it.
- ☆ Take a fresh sample and add 40 per cent sulphuric acid into it, if it get dissolved into it, it confirms soluble nylon and insoluble material might be polyester.
- ☆ The fresh sample can be treated with cyclohexane, the solubility of the material confirms PVC.

4. Identification of Packaging Films by Flame Test

Burning behavior of plastic packaging films and the color and other characteristics of flame can guide us to their tentative identification which can be further confirmed on the basis of chemical testes.

The pieces of films are burnt on a Bunsen flame for a short while and the burning characteristics–the flame, film response, smell etc. are carefully recorded. The following table can be of much help in this respect.

Burning Characteristics of Packaging Films

Plastic Film	*Characteristics of the Flame*	*Burning Behavior of the Film*	*Smell*	*Self extingui-shing*
Polyethylene	Yellow top with blue bottom and some white smoke	Melts and drips	Burnt paraffin wax	No
Polypropylene	-do-	-do-	Acrid, burnt wax	No
Polyvinyl chloride	Orange yellow with green edges	Darkness rapidly, softens and decomposes	Typical chloride	Yes
Polyamide (nylon)	Yellow top with blue bottom	Melts and drips with froths. Drips can not be crushed	Burning hair	Yes
Polyester	Burns steadily with yellow black smoke	No drip	Pleasant resin odour	No
Polystyrene	Orange yellow with black sooty smoke	Softens, without drip	Marigold odour	No
Cellulose acetate	Yellow top with blue bottom	Melts, burns quickly with irregular charred beads	Burnt vinegar	No

5. Estimation of Water Vapour Transmission Rate (WVTR)

Moisture loss or gain, through the packaging film or package during storage and distribution affects the quality and shelf life of food. WVTR is measured as the quantity of water vapour in grams which will permeate through one square meter of a film in 24 hours at 38°C and 90 per cent RH. Thus, WVTR determines the efficiency of a packaging film to the transmission of water vapour. Films and laminates of very low WVTR are used for the packaging of hygroscopic foods.

Reagents and Materials

1. Aluminium Dishes
2. Calcium chloride
3. Micro-crystalline wax for sealing
4. Weighing balance
5. Humidity cabinet

Procedure

- WVTR is measured with the help of specially designed aluminum dish which is filled with finely ground calcium chloride at the bottom.
- Accurately weighed plastic film is put over the dish and all the sides are sealed with microcrystalline wax by using a wax applicator.
- The initial weight of dish is noted.
- The dish is put in the humidity cabinet at a temperature of 38°C and 90 per cent RH.
- The increase in weight of the dish is noted every day for six days.
- Average gain in weight of aluminum dish in 24 hours gives the quantity of water vapour which could permeate through the film.
- The same is calculated on the basis of square meter area of the film. The observations are taken in duplicate.

6. Determination of Gas Transmission Rate (GTR)

Gas transmission rate is measured as the effectiveness of a packaging film in resisting the permeability of a particular gas through a unit area. The measurement is done either in terms of change in volume at constant pressure (usually 1 atmosphere) or in terms of change in pressure at constant volume. At normal temperature and pressure (NTP) it is calculated as follows:

$$\text{GTR} = \frac{24 \times \text{Volume (76)}}{\text{Area (sq.m)} \times \text{Time (hr)} \times \text{Pressure diff. (cm of Hg)}}$$

It is expressed in the unit of $cc/m^2/24$ hr

GTR of a packaging film is different for oxygen or carbon dioxide or other gases. Films of low oxygen transmission rate (OTR) or oxygen barrier laminates are selected for oxygen sensitive foods to prevent the development of oxidative rancidity.

Appendix

Sorenson: Meth Enzymol. 1, 143 (1955)

Na_2HPO_4. 2 H_2O M.Wt. 178.05; 0.2M solution contains 35.61 g/l

Na_2HPO_4. 12 H_2O M.Wt. 358.22.; 0.2M solution contains 71.64 g/l

NaH_2PO_4. H_2O M.Wt. 138.01; 0.2M solution contains 27.6 g/l

NaH_2PO_4. 2 H_2O M.Wt. 156.03; 0.2M solution contains 31.21 g/l

X ml 0.2 M $Na_2HPO_{4,}$ Y ml 0.2 M Na HPO_4; diluted to 100 ml with H_2O to get 0.1 M solution

pH, 25° C	*X ml 0.2M Na_2HPO_4*	*Y ml 0.2M $NaHPO_4$*
5.8	4.0	46.0
6.0	6.15	43.85
6.2	9.25	40.75
6.4	13.25	36.75
6.6	18.75	31.25
6.8	24.5	25.5
7.0	30.5	19.5
7.2	36.0	14.0
7.4	40.5	9.5
7.6	43.5	6.5
7.8	45.75	4.25
8.0	47.35	2.65

Metric Constants

1 kilogram (kgm)	1,000.0 g	2.2046 pounds
1 g	1000mg	15.4 grain
1 l	1000.0 ml	0.21998 gallons
1 ml	0.001 l	0.0352 ounce
1 quintal	100 kg	220.46 pounds
1 pound	12 ounce	5,760 grains
1 barrel	224 pounds	28 close
1 inch	2.54 cm	25.4 mm
1 m	100 cm	39.37 inch
1 kg/cm^2	14.223 p.s.i	–
1 °F	0.555 °C	–

Factors for conversion from one scale to the other

To convert Centigrade into Fahrenheit	multiply by 1.8 and add 32
To convert Fahrenheit into Centigrade	subtract 32 and divide by 1.8
To convert Grams into Grains	multiply by 15.432
To convert Grains into Grams	multiply by 0.0648
To convert Grams into Ounce	multiply by 0.03527
To convert Ounce into Grams	multiply by 28.35
To convert Kilograms into Pounds	multiply by 2.2046
To convert Pounds into Kilograms	multiply by 0.454
To convert Meters into Inches	multiply by 32.37
To convert Inches into Meters	multiply by 0.0254

Relationship Between Transmittance in Per cent (per cent T) and the Absorbance (Extinction of optical density)

Per cent T	A	Per cent T	A	Per cent T	A	Per cent T	A
100	0.000	75	0.125	50	0.301	25	0.602
99	0.004	74	0.131	49	0.310	24	0.620
98	0.009	73	0.137	48	0.319	23	0.638
97	0.013	72	0.143	47	0.328	22	0.658
96	0.018	71	0.149	46	0.337	21	0.678
95	0.022	70	0.155	45	0.347	20	0.699
94	0.027	69	0.161	44	0.357	19	0.721
93	0.032	68	0.168	43	0.367	18	0.745
92	0.036	67	0.174	42	0.377	17	0.770
91	0.041	66	0.171	41	0.387	16	0.796
90	0.046	65	0.187	40	0.398	15	0.824
89	0.051	64	0.194	39	0.409	14	0.854
88	0.056	63	0.201	38	0.420	13	0.886
87	0.061	62	0.208	37	0.432	12	0.921
86	0.066	61	0.215	36	0.444	11	0.959
85	0.071	60	0.222	35	0.456	10	1.000
84	0.076	59	0.229	34	0.469	9	1.046
83	0.081	58	0.237	33	0.482	8	1.097
82	0.086	57	0.244	32	0.495	7	1.155
81	0.092	56	0.252	31	0.509	6	1.222
80	0.097	55	0.260	30	0.523	5	1.301
79	0.102	54	0.268	29	0.538	4	1.398
78	0.108	53	0.276	28	0.552	3	1.523
77	0.114	52	0.284	27	0.569	2	1.699
76	0.119	51	0.292	26	0.585	1	2.000

References

Adler- Nissen J. (1979). Determination of the degree of hydrolysis of food protein hydrolysates by trinitrobenzene sulfonic acid. J.Agriculture and Food Chemistry 27:1256-1261

Akeson, W.R. and Stahmann, M.A. (1964). Pepsin-pancreatin digest complex of protein quality evaluation. J Nutrition. 83 : 257-259

Anand, S. K. (1988). Methods for rapid detection of micro-organisms in meat and meat products. Poultry Guide XXVI (1):72-73

Anderson R A, Conway H F, Pfeifer V F, and Jr. E L. 1969. Gelatinization of corn grits by roll and extrusion-cooking. *Cereal Science Today* 14: 4-12

Anon. (1984). ISI Hand Book of Food Analysis, Sp : 18 (Part XII)

AOAC (1975). *Official Methods of Analysis.* Association of Official Analytical Chemists, 12th ed. Washington, D.C.

AOAC (1980). *Official Methods of Analysis*. 13 th Ed. Association of Official Analytical Chemists. Washington, DC

AOAC (1984) *Official Methods of Analysis* 14th ed. Association of Official Analytical Chemists. Washington, DC

AOAC (1995). *Official Methods of Analysis*. 16th edn. Association of Official Analytical Chemists, Washington, D.C.

APHA (1984). *Compendium of Methods for the Microbiological Examination of Foods.* 2nd edn. (ed. M.L. Speck). American Public Health Association, Washington, D.C.

Arganosa F.C. and Henrickson, R.L. (1969). Cure diffusion through pre- and post-chilled porcine muscles. Food Technology 23: 1061-1065

Baliga, B.R. and Madaiach, N. (1970). Quality of sausages emulsion prepared from mutton. J Food Science 35:383-385

Barker, S.B. and Summerson, W.H. (1941). The colorimetric determination of lactic acid in biological material. J. Biological Chemistry. 138: 535-554

Bater, B, Descamps O. and Maurer, A.J. (1992). Quality characteristics of hydrocolloid-added oven-roasted turkey breasts. J. Food Science 57(5): 1068-1070

Bender, AE and Doell, B.H. (1957). Biological evaluation of proteins. A new aspect. British J Nutrition. 11:140-142

Bernthal, P.H, Booren, A.M and Gray, J.I. (1991) Effect of reduced Sodium Chloride concentration and tetra sodium pyrophosphate on pH, water holding capacity and extractable protein on pre-rigor and post-rigor ground beef. Meat Science 29(1): 69-82

Berry, B.W. and Stiffler, D. M. (1981) Effect of electrical stimulation, boning temperature, formulation and rate of freezing on sensory, cooking, chemical and physical properties of ground beef patties. J Food Science 46:1103-1106

BIS (1989). Handbook of Food Analysis Part XI. Dairy Products. Bureau of Indian Standards. Manak Bhawan, New Delhi.

Booth V. H. (1971) Problems in the determination of FDNB-available lysine. J. Science Food Agriculture 1971; 22:658-666

Bourne, M.C. (1978). Texture profile analysis, Food Technology, 32(7): 62-66, 72.

Burn (1971). Cited from Effect of domestic processing, cooking and germination on the trypsin inhibitor activity and tannin content of faba bean (*Vicia faba*). Plant Foods for Human Nutrition, 1992: 42 (2) 127-133

Cambero, M. I., Seuss, I. and Honikel, K. O. (1992). Flavour components of beef broth as affected by cooking temperature. J Food Science 57(6): 1283-90

Cambero, M. I., Seuss, I. and Honikel, K. O. (1992). Flavour components of beef broth as affected by cooking temperature. J Food Science 57(6): 1283-90

Carpenter, K.J. (1960) The estimation of available lysine in animal protein food. Biochemistry J. 77:604

Chapmen, D.G., Castillo, R. and Campbell, J.A.(1959). Evaluation of protein in foods. Canadian J Biochemistry Physiology 37 : 679

Cheng K. L. and Bray R.H. (1951). Diagnosis and improvement of saline and alkali soils. Agriculture Handbook No. 60. *United States Deptt. of Agril*, p. 94-95.

Chick, H., Hutchinson, J.C.D. and Jackson, H.M.(1935). The biological value of protein VI. Further investigation of balance sheet method. Biochemistry J. 29 :1702-1711

Chu, Y.H., Huffman, D.L., Trout, G.R. and Egbert, W.R. (1987). Color and color stability of frozen restructured beef steak- Effect of sodium chloride, tripolyphosphates, nitrogen atmosphere and processing procedure. J Food Science 52: 869-875.

Corper, G.R. (1976). Standard Method of Clinical Chemistry (Ed), New York Vol. 7. Clin. Chem. 22, 616

Cross, H.R., West, R.L. and Dutson, T.R. (1980). Comparisons for measuring Sarcomere length in beef semitendinosus muscle. Meat Science, 5: 261-266

Culler R. D., Parrish F. C, Smith G. C. and Cross H. R. (1978). Relationship of myofibril fragmentation index to certain chemical, physical and sensory characteristics of bovine longissimus muscle. J. of Food Science, 43 (4):1177-1180

Davies N. T.and Reid H. (1979). An evaluation of the phytate, zinc, copper, iron and manganese contents of zinc availability from soya-based textured-vegetable-protein-meat-substitutes or meat extenders. British J. Nutrition; 41:579-589

Davis, G.W., Dutson, T.R., Smith, G.C. and Carpenter, Z.L. (1980). Fragmentation procedure for bivine longissimus muscle as an index of cooked steak tenderness. J of Food Science, 45: 880-885

Dierick, N., Vandekerckhove, P. and Demeyer, D. (1974). Changes in non-protein nitrogen compounds during dry sausage ripening. J. Food Science. 39: 301–304.

Drabkin D.L. (1950) The distribution of the chromoproteins, hemoglobin, myoglobin, and cytochrome c, in the tissues of different species, and the relationship of the total content of each chromoprotein to body mass. J. Biol Chem. 182: 317

Everson, C.W., Keyahian, T. and Doty, D.M. (1955). Fat and moisture: rapid methods for determination of fat and moisture content in meat products. American Meat Institute Foundation Bulletin No. 26

Fisher K (1935) A new method for the analytical determination of the water content of liquids and solids. Angew Chem 48: 394-396

Fiske, C.H and Sabbarow, Y. (1925). Colorimetric determination of phosphorus J. Biological Chemistry. 66: 375

Folch, J., Lees, M. and Sloane-Stanley, G.H. (1957). A simple method for the isolation and purification of total lipids from animal tissues. J. Biological Chemistry 226: 497-509

Froehlich, D.A., Gullet, E.A. and Usborn, W.R. (1983).Effect of nitrite and salt on the color, flavor and overall acceptability of ham. J. of Food Sci. 48(1): 152-154.

Gardener, G.A. and Stewart, D. J. (1966). Changes in free amino acids and other nitrogen compounds in stored beef muscles. J Science Food and Agriculture. 17: 491-496

Gestetnev, B., Brik, Y., Bondi, A. and Tencer, Y., (1966). Method for the determination of sapogenin and saponin contents in soybeans. Phytochemistry 5 : 803-806.

Gnanasambandam, R and Zayas, J. F. (1992) Functionality of wheat germ protein in comminuted meat products as compared with corn germ and soy proteins. J Food Science 57(4): 829-833

Goll, D.E., Bray, R.W. and Hoekstra, W.G. (1963). Age-associated changes in muscle composition- The isolation and properties of a collagenous residue from bovine muscle. J Food Science, 28: 503

Gram, L. and Sagaard, H. (1985). Microcalorimetry as a rapid method for estimation of bacterial levels in ground meat. J. Food Prot. 48: 341-345.

Grau, R. and Hamm, R. (1957). Uber das Wasserbindungsvermo¨ gen des Sa¨ugetiermuskels. Z. Lebensm.Unters. Forsch 105: 446–460.

Grau, R., and Hamm, R. (1953). Eine enfache Methode zur Bestimmung der Wasserbindung im Mukel. *Naturwissenschaften* 40: 29.

Gupta, H.O., Lodha, M.L., Mehta, S.I., Rastogi, D.K. and Singh, J. (1979). Effect of amino acids and pulse supplementation on nutritional quality of normal and modified opaque-2 maize (*Zea mays* L.). J Agriculture Food Chemistry 27 (4) :787-790

Hall, R.T., Trienders N. and Givens, D.L. (1973). Observations on the use of 2,4,6- trinitrobenzene sulphonic acid for the determination of available lysine in animal protein concentrates. Analyst, 98 : 673-686.

Hamm, R. (1960) Biochemistry and meat hydration. Advances Food Research, 10: 363

Hanel H.K. and Dam H. (1955). Determination of small amount of total cholesterol by Tschugaeff reaction with a note on the determination of lanosterol. Acta Chem Scandinivian 9: 677-682

Haque, N. and Murrari Lal. (1999). Gross energy estimation. In: Laboratory Manual of Animal Nutrition. Centre of Advanced Studies, Division of Animal Nutrition, Indian Veterinary Research Institute, Izatnagar, U.P, India. Pp. 71-76.

Hegarty G.R., Bratzler, L.J. and Pearson, A.M. (1963). The relationship of some intracellular protein characteristics to beef muscle tenderness. J Food Sci. 28:525

Hernandez-Ledesma, B, Beatriz Miralles, Lourdes Amigo, Mercedes Ramos and Isidra Recio., 2005 Identification of antioxidant and ACE-inhibitory peptides in fermented milk, *J Sci Food Agric.,* 85:1041-1048.

Hill, F. (1966). The solubility of intramuscular collagen in meat animals of various ages. J. Food Science. 31: 161

Holland, D.C. (1971). Determination of malonaldehyde as an index of rancidity in nut meats. J Assoc. Official Analytical Chemists. 54, 1024

Honikal, K.O. (1997) Reference methods supported by OECD and their use in Mediterranean meat products. Food Chemistry. 59:573-582

Honikal, K.O. and Fischer, C. (1977). A rapid method for the detection of PSE and DFD porcine muscles. J Food Science 42, 1633-36.

Hornsey, H.C. (1956). The color of cooked pork. I. Estimation of the nitric oxide haeme pigments. Journal of the Science of Food and Agriculture, 7: 534-540

Hostettler, F., Borel, E. and Deuel, H. (1951). Uber die Reduction der 3,5-Dinitrosalicylsaure durch Zucker. (Reduction of 3,5-dinitrosalicylic acid by sugars). Helvetica Chimica Acta 34, 2132-2139.

Ibrahim, G.F. (1986) A review of immunoassays and their application to salmonellae detection in foods. J. Food Protection 49, 299-310

Ingram, M. and Dainty, R.H. (1971) Changes caused by microbes in spoilage of meat. J. Applied Bacteriology 34:21-39

International Organization for Standardization (1974). Meat and meat products- determination of hydroxyproline content. Draft International Standard, ISO/DIS 3496

Ishikawa J., Fuse, Y. and Wakabayashi, K. (1987). Choice of extraction procedure for estimation of anterior pituitary hormone content. Endocrinology Japanese 34 : 21-29

ISI (1961). IS: 1479. Indian Standard methods of tests for dairy industry. Part II. Chemical analysis of milk. Indian Standards Institution, Manak Bhawan, New Delhi.

ISI (1966) IS : 3508. Indian Standard method of sampling and test for ghee (butter fat) Indian Standards Institution, Manak Bhawan, New Delhi.

ISI (1967) IS: 1165 (Revised). Specification for milk powder (whole and skim), Indian Standard Institution, Manak Bhawan, New Delhi.

ISI (1984) Estimation of moisture. Sp : 18 (Part XII)–1984. ISI Hand Book of Food Analysis pp 13

Jackson, M.L. (1962). Soil Chemical Analysis. Asia Publishing House, P 151-154.

James, N.A. and Ryley, J. (1986). The rapid determination of chemically reactive lysine in the presence of carbohydrates by a modified trinitrobenzene sulphuric acid procedure. J. Science Food Agriculture,37 (2) : 1551-56.

Jay, J. M. (1964). Beef microbial quality determined by extract release volume (ERV). Food Technology 18: 1637-1641

Jay, J.M. (1977). The Limulus lysate endotoxin assay as a test of microbial quality of ground beef. J. Applied Bacteriology 43:99-109

Jay, J.M. (1987) Meats, Poultry and Seafoods. In Food and Beverage Mycology, 2nd ed. Ed. LR Beuchat, Chap 5. New York: Kluwner Academic Publisher

Jeremiah, L.E. and Martin, A.H. (1982). Effect of prerigor chilling and freezing and subcutaneous fat cover upon the histological and shear properties of bovine longissimus dorsi muscle. Journal of Animal Science, 62: 353-361

Juncher, D., Vestergaard, C.S., Soltoft-Jensen, J., Weber, C.J., Bertelsen, G. and Skibsted, L.H. (2000). Effects of chemical hurdles on microbiological and oxidative stability of a cooked cured emulsion type meat products. Meat Science, 55: 483-491

Kang, C.K. and Rice E.E. (1970). Degradation of various meat fractions by tenderizing enzymes. J Food Science. 35:563.

Karl, D.M. (1980) Cellular nucleotide measurements and applications in microbial ecology. Microbiological Review,44; 739-396

Ke, P.J., Cervantes, E. and Robles–Martinez, C. (1984). Determination of Thiobarbituric Acid Reactive Substances (TBARS) in fish tissue by an improved distillation–Spectrophotometric method. J. Sci Food Agric, 35 (11) : 1248-1254.

Kennedy, J.E and Oblinger, J.L (1985). Application of bioluminescence to rapid determination of microbial levels in ground beef. J. Food Protection 48: 334-340

Knipe, C.L., Olson, D.G. and Rust, R.E. (1985). Effect of selected inorganic phosphates, phosphate levels and reduced sodium chloride levels on protein solubility, stability and pH of meat emulsion. J Food Sci. 50(4) :1010-1013

Koniecko, E.S. (1979). In: Handbook for Meat Chemists. Chap. 6, Avery Publishing Group Inc., Wayne, New Jersey, pp 68-69

Koniecko, E.S. (1979). Handbook for meat chemists, Avery Publishing group, Inc. Wayne, New Jersey pp. 53- 55.

Kramlich, W.E., Pearson, A.M. and Tauber F.W. (1973) Processed Meats. AVI Pub. Co.

Krzywicki, K. (1979). Assessment of relative content of myoglobin, oxymyoglobin and metmyoglobin at the surface of meat. Meat Sci. 3:1-10

Labuza, T.P., Tannenbaum, S.R. and Karl, M. (1970). Water content and stability of low moisture and intermediate moisture foods. Food Technol. 24: 543.

Laemmli, U.K. (1970). Cleavage of structural proteins during the assembly of the head of bacteriophage T_4. Nature 227:680-685

Landrock, A.H. and Proctor, BE (1951a) A new graphical interpolation method for obtaining humidity equilibria data. Food Tech, 332.

Landrock, A.H. and Proctor, B.E. (1951b) Measuring humidity equilibria. Modern Packaging 24, 123-130, 186.

Little, A.C. (1975) Off on a tangent. J Food Sci. 40(2):410-412.

Liu, F., Ooi, V.E.C. and Chang, S.T. 1997. Free radical scavenging activities of mushroom polysaccharide extracts, *Life Sci.*, 60(10): 763-771

Lowry, O.H., Rosenbrough, N., Farr, A.L. and Randall, R.J. (1951). Protein measurement with the Folin Phenol reagent. J Biol. Chem. 193:265-275

Machado, M.F., Oliveria, F.A.R., Gekas, V. and Singh, R.P., 1998 Kinetics of moisture uptake and soluble solids loss by puffed breakfast cereals immersed in water. *International J. Food Sci. and Technology* 33: 225-237.

Mangold, H.K. (1984) ed. CRC Handbook of Chromatography; Lipids, Vol. 1, Boca Raton, FL: CRC Press, pp 387-397

Marinetti, G.V. 1962. Chromatographic separation, identification and analysis of phosphatides. J Lipid Res. 3: 1-20

Mittal, P. and Lawrie, R.A. 1986. Extrusion studies of mixtures containing certain meat offals. Part 2. Textural properties. *Meat science* 16: 143-160.

Neuman, R.E. and Logan, M.A. (1950). The determination of hydroxyproline. Journal of Biological Chemistry, 184: 299-306

Notermans, S. Intveld, P., Wijtzes, T. and Mead, G.C (1993). A user's guide to microbial challenge testing for ensuring the safety and stability of food products. Food Microbiology 10:145

Ockerman, H.W. (1976). Quality control of postmortem muscle tissue 10th ed. vol.1 The OSU, Deptt. of Anim. Sci. Columbus, Ohio.

Okonkwo, T.M., Obanu, Z.A. and Ledward, D.A. (1992). Characteristics of intermediate moisture smoked meats. Meat Science, 31(22): 135-145

Olson, B.R., Guzman, N.A., Engel, J., Condt, C. and Steiner, A. (1977). Purification and characterization of a peptide from carboxy terminal region of chick tender procollagen type I. Biochemistry, 16 (13) 3030

Parks, L.L. and Carpenter, J.A. (1987). Functionality of six non-meat proteins in meat emulsion systems. Journal of Food Science. 52 (2) :271-274

Pearson, D. (1968). Application of chemical methods for the assessment of beef quality II. Methods related to protein breakdown. J. Science Food Agriculture 19: 366-369

Pellegrini, N., Re, R., Yang, M., and Rice-Evans, C. (1999). Screening of dieatary carotenoids and carotenoid-rich fruit extracts for antioxidant activities applying 2,2′-azinobis (3-ethylenebenzothiazolin-6-sulpfonic acid) radical cation decolorization assay. *Methods Enzymol.*, 299:379-389.

Pettipher, G.L and Rodrigues, U.M. (1982). Rapid enumeration of microorganisms in foods by the direct epifluorescent filter technique. Applied Environmental Microbiology 44: 809-813

Pierce K.N. and Kinsella, J.E. (1978). Emulsifying properties of proteins: Evaluation of a turbidimetric technique. J Agriculture Food Chemistry 26, 716–723.

Platt, B.S., Miller, D.S. and Payne, P.R. (1961). Protein values of human foods. In : Recent advances in human nutrition. Ed. Brock, J.F., London, pp. 351.

Raghuramulu, N., Nair, K.M. and Kalyanasundaram, S. (1983a). Carbohydrates : Glucose- glucose oxidase method (J Clinical Pathology(1969) 22, 246) A manual of laboratory Techniques, published by National Institute of Nutrition, Hyderabad, pp. 73-74.

Raghuramulu, N., Nair, K.M. and Kalyanasundaram, S. (1983b). Carbohydrates : Glucose- Ferricyanide method (Methods in Enzymology(1957) vol III, 86) A Manual of Laboratory Techniques, published by National Institute of Nutrition, Hyderabad, pp. 74-75.

Raghuramulu, N., Nair, K.M. and Kalyanasundaram, S. (1983c). Carbohydrates : Glucose- Hagedorn and Jensen method (Howk, Oser and Summerson (1954) Practical Physiological Chemistry, Eds 13th (Edn.) 577) A Manual of Laboratory Techniques, published by National Institute of Nutrition, Hyderabad, pp. 75-76.

Raghuramulu, N., Nair, K.M. and Kalyanasundaram, S. (1983d). Carbohydrates : Glucose- Nelson and Somogyl method (Methods in Enzymology (1957) vol III, 87) A Manual of Laboratory Techniques, published by National Institute of Nutrition, Hyderabad, pp. 77-78.

Raghuramulu, N., Nair, K.M. and Kalyanasundaram, S. (1983e). Carbohydrates : Liver Glycogen (J Biological Chemistry (1940) 135, 511) A Manual of Laboratory Techniques, published by National Institute of Nutrition, Hyderabad, pp. 78.

Rajkumar, V., Agnihotri, M.K. and Sharma, N. (2004). Quality and shelf life of vacuum and aerobic packed chevon patties under refrigeration. Asian Australasian J. of Animal Science 17(4): 548-553.

Rosen, H. (1957). A modified ninhydrin colorimetric analysis for amino acids. Arch Biochemical Biophysics, 67(9), 10-15.

Roughan, P.G. and Batt, R.D. (1968). Quantitative analysis of sulfolipid (sulfoquinoxosyl diglyceride) and galactolipids (monogalactosyl and digalactosyl diglycerides) in plant tissues. Anal Biochem 22: 74-88.

Rowland, S.J. (1938). The determination of nitrogen distribution in milk. J. Dairy Science. 9 42–47

Saffle, R.L and Galbreath, J.W (1964). Quantitative determination of salt soluble protein in various types of meat. Food Technology 18, 1943.

Sahoo, J.(1995). Effect of preblending and vacuum packaging on the quality of ground buffalo meat. Ph.D. thesis, Deemed University Indian Veterinary Research Institute, Izatnagar.

Salih, A.M, Smith, D.M and Dawson, L.E. (1987). Modified extraction of 2-thiobarbitaric acid method for measuring lipid oxidation in poultry. Poultry Science 66: 1483.

Sempore, G. and Bezard, J. (1991). Determination of molecular species of oil triacylglycerols by reversed-phase and chiral-phase high-performance liquid chromatography. J American Oil Chemists Society. 68:702-709

Sempore, G. and Bezard, J. (1977). Fatty acid characterization–fat (Folch procedure).Rev Fr Corps Gras 24:611

Shelf and Jay (1970). Use of titrimetric method to assess the bacterial spoilage of fresh meats. Bact. Proc. 70:6

Shimada, K., Fujikawa, K., Yahara, K. and Nakamura, T. 1992. Antioxidative properties of xanthane on the autoxidation of soybean oil in cyclodextrin emulsion. *J. Agric. Food Chem.*, 40: 945-948

Shivas, S.D., Kropf, D.H., Hunt, M.C., Kastner, C.L., Kendal, J.L.A. and Dayton, A.D. (1984). Effect of ascorbic acid on display life of ground beef. J. Food Protect.47(1)1-15.

Singh, V. and Jambunathan, R. (1981). Studies on desi and kabuli chickpea (*Cicer arietinum* L.) cultivar levels of protease inhibitor level of polyphenolic compounds and in vitro digestibility. J. Food Sci. 46: 1364-1367

Standard, C.J and Wood, J.M, (1983) The rapid estimation of microbial contamination of raw meat by measurement of adenosine triphosphate (ATP). J. Applied Bacteriolology 55: 429-438

Stegman, H. and Stadler, K. (1967). Determination of hydroxyproline. Clin. Chem. Acta. 18:267.

Swift, C.E., Locker, C. and Fryar, A.J. (1961). Comminuted meat emulsion–the capacity of meat for emulsifying fat. Food Technology, 15: 468-473.

Tarladgis, B.G., Watts, B.M., Younathan, M.T. And Dugan, L.R. (1960). A distillation method for the quantitative determination of malonaldehyde in rancid foods. J. American Oil Chemists Society, 37, 44-48.

Townsend, W.E., Witnauer, L.P., Riloff, J.A. and Swift, C.E. (1968). Communited meat emulsion: Differential thermal analysis of fat transition. Food Technol. 22: 319-323.

Trout, E.S., Hunt, M.C., Johnson, D.E., Claus, J.R., Kastner, C.L., Kropf, D.H. and Stroda, S. (1992). Chemical, physical and sensory characterization of ground beef containing 5 to 30 per cent fat. J Food Science, 57:25-29.

Trout, G.R. (1989) Variations in myoglobin denaturation and color of cooked beef, pork and turkey meat as influenced by pH, sodium chloride, sodium tripolyphosphate and cooking temperature.J. Food Science. 54(3): 536-540.

Tuma, H.J., Henrickson, R.L., Stephens, D.F. and Moore, Ruby (1962) Influence of marbling and animal age on factors associated with beef quality. J. Animal Science 21, 848-851.

Wardlow, F.R., Mc Caskill, L.H. and Acton, J.C. (1973). Effects of post mortem changes on poultry meat loaf properties. J Food Science 38: 421-423

Warris, P.D. (1979). The extraction of heme pigments from fresh meat. J. Food Technology 14:75.

Webb, N.B., Ivey, J.F., Craig, H.B., Jones, V.A. and Monroci, R.J (1970). The measurement of emulsifying capacity by electrical resistance. J. Food Sci. 35:501-504.

Wierbicki, E. and Deatherage, F.E. (1958). Determination of water holding capacity of fresh meats. J. Agriculture Food Chemistry, 6: 387.

Witte, V.C., Krause, G.F. and Bailey, M.E. (1970). A new extraction method for determining 2-thiobarbituric acid value of pork and beef during storage. J. Food Science 35: 582-585.

Woessner, J.F. (1961). The determination of hydroxyproline in tissue and protein samples containing small proportions of amino acid. Archives of Biochemistry and biophysics, 93: 440-442.

Zlatkis, A., Zak, B., Boyle, H.J. and Mich, D. (1953). A new method for direct determination of serum cholesterol. J. Laboratory Clinical Medicine, 41: 486-936.

Index